Algebra 1

Study Notebook

Glencoe McGraw-Hill

McGraw Hill Education

connectED.mcgraw-hill.com

Education

Copyright © 2012 The McGraw-Hill Companies, Inc.

All rights reserved. The contents, or parts thereof, may be reproduced in print form for non-profit educational use with Glencoe *Study Notebook: Algebra 1*, provided such reproductions bear copyright notice, but may not be reproduced in any form for any other purpose without the prior written consent of The McGraw-Hill Companies, Inc., including, but not limited to, network storage or transmission, or broadcast for distance learning.

Send all inquiries to:
McGraw-Hill Education
8787 Orion Place
Columbus, OH 43240

ISBN: 978-0-07-660287-2
MHID: 0-07-660287-7

Printed in the United States of America.

5 6 7 8 9 10 RHR 16 15 14 13 12

The McGraw-Hill Companies

Contents

Chapter 1
Before You Read 1
Key Points 2
1-1 Variables and Expressions..................... 3
1-2 Order of Operations............................... 5
1-3 Properties of Numbers 7
1-4 The Distributive Property....................... 9
1-5 Equations... 11
1-6 Relations.. 13
1-7 Functions... 15
1-8 Interpreting Graphs of Functions............. 17
Tie It Together... 19
Before the Test 20

Chapter 2
Before You Read 21
Key Points 22
2-1 Writing Equations 23
2-2 Solving One-Step Equations 25
2-3 Solving Multi-Step Equations 27
2-4 Solving Equations with the Variable
 on Each Side 29
2-5 Solving Equations Involving
 Absolute Value 31
2-6 Ratios and Proportions......................... 33
2-7 Percent of Change 35
2-8 Literal Equations and
 Dimensional Analysis 37
2-9 Weighted Averages 39
Tie It Together... 41
Before the Test 42

Chapter 3
Before You Read 43
Key Points 44
3-1 Graphing Linear Equations..................... 45
3-2 Solving Linear Equations
 by Graphing.. 47
3-3 Rate of Change and Slope..................... 49
3-4 Direct Variation..................................... 51
3-5 Arithmetic Sequences as
 Linear Functions................................... 53
3-6 Proportional and Nonproportional
 Relationships.. 55
Tie It Together... 57
Before the Test 58

Chapter 4
Before You Read 59
Key Points 60
4-1 Graphing Equations in Slope-Intercept
 Form 61
4-2 Writing Equations in Slope-Intercept
 Form 63
4-3 Writing Equations in Point-Slope
 Form 65
4-4 Parallel and Perpendicular Lines 67
4-5 Scatter Plots and Lines
 of Fit 69
4-6 Regression and Median-Fit Lines 71
4-7 Inverse Linear Functions 73
Tie It Together... 75
Before the Test 76

Chapter 5
Before You Read 77
Key Points 78
5-1 Solving Inequalities by Addition and
 Subtraction 79
5-2 Solving Inequalities by Multiplication
 and Division... 81
5-3 Solving Multi-Step Inequalities 83
5-4 Solving Compound Inequalities............. 85
5-5 Inequalities Involving Absolute
 Value 87
5-6 Graphing Inequalities in Two
 Variables... 89
Tie It Together... 91
Before the Test 92

Chapter 6
Before You Read 93
Key Points 94
6-1 Graphing Systems of Equations.............. 95
6-2 Substitution 97
6-3 Elimination Using Addition and
 Subtraction 99
6-4 Elimination Using Multiplication........... 101
6-5 Applying Systems of Linear
 Equations.. 103
6-6 Systems of Inequalities 105
Tie It Together... 107
Before the Test 108

Chapter 7
- Before You Read 109
- Key Points 110
- 7-1 Multiplication Properties of Exponents ... 111
- 7-2 Division Properties of Exponents 113
- 7-3 Rational Exponents 115
- 7-4 Scientific Notation 117
- 7-5 Exponential Functions 119
- 7-6 Growth and Decay 121
- 7-7 Geometric Sequences as Exponential Functions 123
- 7-8 Recursive Formulas 125
- Tie It Together 127
- Before the Test 128

Chapter 8
- Before You Read 129
- Key Points 130
- 8-1 Adding and Subtracting Polynomials 131
- 8-2 Multiplying a Polynomial by a Monomial 133
- 8-3 Multiplying Polynomials 135
- 8-4 Special Products 137
- 8-5 Using the Distributive Property 139
- 8-6 Solving $x^2 + bx + c = 0$ 141
- 8-7 Solving $ax^2 + bx + c = 0$ 143
- 8-8 Differences of Squares 145
- 8-9 Perfect Squares 147
- Tie It Together 149
- Before the Test 150

Chapter 9
- Before You Read 151
- Key Points 152
- 9-1 Graphing Quadratic Functions 153
- 9-2 Solving Quadratic Equations by Graphing 155
- 9-3 Transformations of Quadratic Functions 157
- 9-4 Solving Quadratic Equations by Completing the Square 159
- 9-5 Solving Quadratic Equations by Using the Quadratic Formula 161
- 9-6 Analyzing Functions with Successive Differences 163
- 9-7 Special Functions 165
- Tie It Together 167
- Before the Test 168

Chapter 10
- Before You Read 169
- Key Points 170
- 10-1 Square Root Functions 171
- 10-2 Simplifying Radical Expressions 173
- 10-3 Operations with Radical Expressions 175
- 10-4 Radical Equations 177
- 10-5 The Pythagorean Theorem 179
- 10-6 Trigonometric Ratios 181
- Tie It Together 183
- Before the Test 184

Chapter 11
- Before You Read 185
- Key Points 186
- 11-1 Inverse Variation 187
- 11-2 Rational Functions 189
- 11-3 Simplifying Rational Expressions 191
- 11-4 Multiplying and Dividing Rational Expressions 193
- 11-5 Dividing Polynomials 195
- 11-6 Adding and Subtracting Rational Expressions 197
- 11-7 Mixed Expressions and Complex Fractions 199
- 11-8 Rational Equations 201
- Tie It Together 203
- Before the Test 204

Chapter 12
- Before You Read 205
- Key Points 206
- 12-1 Samples and Studies 207
- 12-2 Statistics and Parameters 209
- 12-3 Distributions of Data 211
- 12-4 Comparing Sets of Data 213
- 12-5 Simulation 215
- 12-6 Permutations and Combinations 217
- 12-7 Probability of Compound Events 219
- 12-8 Probability Distributions 221
- Tie It Together 223
- Before the Test 224

Note-Taking Tips

Your notes are a reminder of what you learned in class. Taking good notes can help you succeed in mathematics. The following tips will help you take better classroom notes.

- Before class, ask what your teacher will be discussing in class. Review mentally what you already know about the concept.
- Be an active listener. Focus on what your teacher is saying. Listen for important concepts. Pay attention to words, examples, and/or diagrams your teacher emphasizes.
- Write your notes as clear and concise as possible. The following symbols and abbreviations may be helpful in your note-taking.

Word or Phrase	Symbol or Abbreviation	Word or Phrase	Symbol or Abbreviation
for example	e.g.	not equal	≠
such as	i.e.	approximately	≈
with	w/	therefore	∴
without	w/o	versus	vs
and	+	angle	∠

- Use a symbol such as a star (★) or an asterisk (∗) to emphasis important concepts. Place a question mark (?) next to anything that you do not understand.
- Ask questions and participate in class discussion.
- Draw and label pictures or diagrams to help clarify a concept.
- When working out an example, write what you are doing to solve the problem next to each step. Be sure to use your own words.
- Review your notes as soon as possible after class. During this time, organize and summarize new concepts and clarify misunderstandings.

Note-Taking Don'ts

- Don't write every word. Concentrate on the main ideas and concepts.
- Don't use someone else's notes as they may not make sense.
- Don't doodle. It distracts you from listening actively.
- Don't lose focus or you will become lost in your note-taking.

NAME _____ DATE _____ PERIOD _____

Expressions, Equations, and Functions

Before You Read

Before you read the chapter, think about what you know about expressions, equations, and functions. List three things you already know about them in the first column. Then list three things you would like to learn about them in the second column.

K What I know…	W What I want to find out…

Foldables Study Organizer Construct the Foldable as directed at the beginning of this chapter.

 Note Taking Tips

- **When taking notes, write down a question mark to anything you do not understand.**

  Before your next quiz, ask your instructor to explain these sections.

- **When you take notes, be sure to listen actively.**

  Always think before you write, but don't get behind in your note-taking. Remember to enter your notes legibly.

Chapter 1   1   Glencoe Algebra 1

NAME _____ DATE _____ PERIOD _____

CHAPTER 1: Expressions, Equations, and Functions

Key Points

Scan the pages in the chapter and write at least one specific fact concerning each lesson. For example, in the lesson on properties of numbers, one fact might be that zero has no reciprocal (because any number times 0 is 0). After completing the chapter, you can use this table to review for your chapter test.

Lesson	Fact
1-1 Variables and Expressions	
1-2 Order of Operations	
1-3 Properties of Numbers	
1-4 The Distributive Property	
1-5 Equations	
1-6 Relations	
1-7 Functions	
1-8 Interpreting Graphs of Functions	

Chapter 1   Glencoe Algebra 1

NAME _____ DATE _____ PERIOD _____

1-1 Variables and Expressions

What You'll Learn

Scan the text under the *Now* heading. List two things you will learn about in this lesson.

1. _____

   _____

2. _____

   _____

Active Vocabulary

New Vocabulary Match each term with its definition.

algebraic expression — the quantities being multiplied in an expression involving multiplication

term — consists of one or more numbers and variables along with one or more arithmetic operations

power — the result of a multiplication expression

factors — symbols used to represent unspecified numbers or values in algebra

product — indicates the number of times the base is used as a factor

variables — a part of an expression that may be a number, a variable, or a product or quotient of numbers and variables

Vocabulary Link *Vary* is a word used in everyday English that is used to build the word *variable*. Find the definition of *vary* using a dictionary. Explain how its everyday definition can help you understand the meaning of *variable* in mathematics.

_____

_____

_____

Chapter 1 3 Glencoe Algebra 1

Lesson 1-1 (continued)

Main Idea	Details
Write Verbal Expressions	Write a verbal expression for each algebraic expression. 1. $4x + 10$ _____ 2. $p - 17$ _____ 3. $\dfrac{3y}{8}$ _____
Write Algebraic Expressions	A model can be used to aid in translating a verbal expression into an algebraic expression. Write an algebraic expression for the following verbal expression. *Twelve more than the product of 8 and h.* 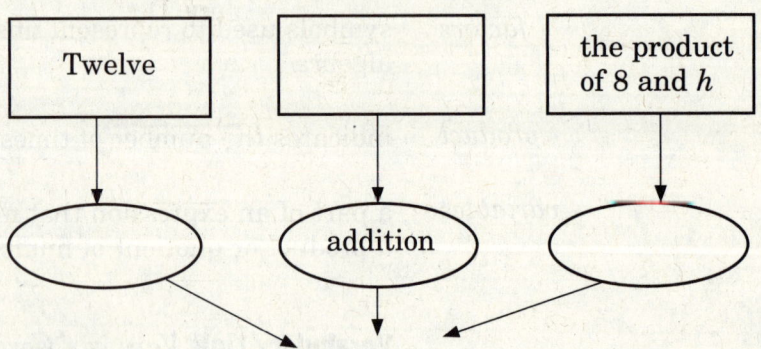

Helping You Remember

A classmate states that 7 less than w translates to $7 - w$. You correct the classmate by saying it translates to $w - 7$. Your classmate responds by saying, "That is the same thing." Is your classmate correct? Support your answer with examples that either disprove or support your classmate.

_____

_____

NAME _____ DATE _____ PERIOD _____

1-2 Order of Operations

What You'll Learn

Skim the lesson. Predict two things that you expect to learn based on the headings and the Key Concept box.

1. _____
   _____
   _____

2. _____
   _____
   _____

Active Vocabulary

Review Vocabulary Write the correct term next to each definition. *(Lesson 1-1)*

_____ ▶ symbols used to represent unspecified numbers or values

_____ ▶ the result of a multiplication expression

_____ ▶ indicates the number of times the base is used as a factor

New Vocabulary Define the following terms in your own words.

evaluate ▶ _____
_____

order of operations ▶ _____
_____

Vocabulary Link *Evaluate* is a word that is used in everyday English. Find the definition of *evaluate* using a dictionary. Explain how its English definition can help you understand its meaning in mathematics.

_____
_____
_____

Chapter 1   5   Glencoe Algebra 1

NAME _____ DATE _____ PERIOD _____

Lesson 1-2 (continued)

Main Idea	Details

Evaluate Numerical Expressions

Evaluate each expression.

1. 3^3 _____

2. $4(2 + 3) - 8$ _____

3. $(4 + 2)^2 \div 2$ _____

Evaluate Algebraic Expressions

Complete the chart that shows the steps in evaluating an algebraic expression.

Replace the _____ with their assigned _____.	Apply the _____ _____ to the expression.	_____ and label your answer, if necessary.

Helping You Remember

Complete each rung of the ladder with the correct order of operations. Start at the bottom and work your way to the top.

- Aunt Sally
- My Dear
- Excuse
- Please

Chapter 1   Glencoe Algebra 1

NAME _____ DATE _____ PERIOD _____

1-3 Properties of Numbers

What You'll Learn

Scan the text in the lesson. Write two facts you learned about properties of numbers as you scanned the text.

1. _____

_____

2. _____

_____

Active Vocabulary

Review Vocabulary Define *variables* in your own words. *(Lesson 1-1)*

_____

_____

New Vocabulary Fill in each blank with the correct term or phrase.

equivalent expressions — Two numbers whose product is 1 are called *multiplicative inverses* or _____.

reciprocals — Expressions that represent the same number are _____.

Additive Identity — The number 1 is known as the _____.

Multiplicative Identity — The number 0 is known as the _____.

Vocabulary Link *Identity* is a word that is used in everyday English. Find the definition of *identity* using a dictionary. Explain how its English definition can help you understand its meaning in mathematics, specifically when referring to additive and multiplicative identities.

_____

_____

_____

Chapter 1   7   Glencoe Algebra 1

NAME _____ DATE _____ PERIOD _____

Lesson 1-3 (continued)

Main Idea	Details

Properties of Equality and Identity

Fill in the blanks with the property used in each step.

$5(9 + 3) \cdot (9 - 8) \cdot \frac{1}{60} + (-5 + 5)$

$= 5(12) \cdot (1) \cdot \frac{1}{60} + (-5 + 5)$ _____

$= 5(12) \cdot (1) \cdot \frac{1}{60} + 0 \quad -5 + 5 = 0$ _____

$= 60 \cdot (1) \cdot \frac{1}{60} + 0 \qquad 5(12) = 60$ _____

$= 60 \cdot \frac{1}{60} + 0 \qquad 60 \cdot 1 = 60$ _____

$= 1 + 0 \qquad 60 \cdot \frac{1}{60} = 1$ _____

$= 1 \qquad 1 + 0 = 1$ _____

Use Commutative and Associate Properties

Use the Associative Property to write two equivalent expressions. Use the numbers 4, 6, and 9.

Use the numbers and a set of parentheses to write an addition expression.

⟶ ⬭ = 19

⟶ ⬭ = 19

Helping You Remember

Look up the meaning of the word *commute* in the dictionary. Find an everyday meaning that is close to the mathematical meaning and explain how it can help you remember the mathematical meaning.

_____

_____

NAME _____ DATE _____ PERIOD _____

1-4 The Distributive Property

What You'll Learn
Scan the lesson. List two headings you would use to make an outline of this lesson.

1. _____

_____

2. _____

_____

Active Vocabulary

Review Vocabulary Write the term next to each definition. *(Lesson 1-2)*

_____ ▶ to find the value of an expression

_____ ▶ the rules that let you know which operation to perform

New Vocabulary In the diagram, underline the *coefficient*.

$$10y + 7$$

Define *simplest form* in your own words.

_____

_____

Vocabulary Link *Distribute* is a word that is used in everyday English. Find the definition of *distribute* using a dictionary. Explain how the English definition can help you remember how *distributive* is used in mathematics.

_____

_____

_____

_____

Chapter 1 — Glencoe Algebra 1

NAME _____ DATE _____ PERIOD _____

Lesson 1-4 *(continued)*

Main Idea	Details
Evaluate Expressions	Caitlin works at the Dairy Whiz Monday through Friday. She earns $8.25 per hour. The hours she worked this week are shown in the table below. Write two equivalent ways of finding her weekly pay.

Day	Mon	Tue	Wed	Thu	Fri
Hours	3	2	$1\frac{1}{2}$	4	$1\frac{1}{2}$

Method 1: hourly rate of pay times total hours for the week

_____

Method 2: hourly rate of pay times daily hours worked

_____

Simplify Expressions   Model the expression $4(x + 3)$ by using or drawing algebra tiles. Then simplify.

Helping You Remember   Write one example of *evaluating an algebraic expression* and explain how you simplified it.

_____

_____

Chapter 1   10   Glencoe Algebra 1

NAME _____ DATE _____ PERIOD _____

1-5 Equations

What You'll Learn Skim the Examples in the lesson. Predict two things you think you will learn about this lesson.

1. _____

_____

_____

2. _____

_____

_____

Active Vocabulary **New Vocabulary** True or False? All open sentences are equations. Explain your answer.

_____

_____

Label the elements of the table with the correct terms.

equation

replacement set

solution

x	3x + 1 = 10	True or False?
2	3(2) + 1 = 10	False
3	3(3) + 1 = 10	True
4	3(4) + 1 = 10	False
5	3(5) + 1 = 10	False

Vocabulary Link In mathematics, *sets* are collections of objects or numbers. *Sets* can be illustrated by real-world examples, like a chess *set*. Write another example of a real-world *set*.

_____

_____

Chapter 1   11   Glencoe Algebra 1

Lesson 1-5 (continued)

Main Idea	Details
Solve Equations	How to solve multi-step linear equations 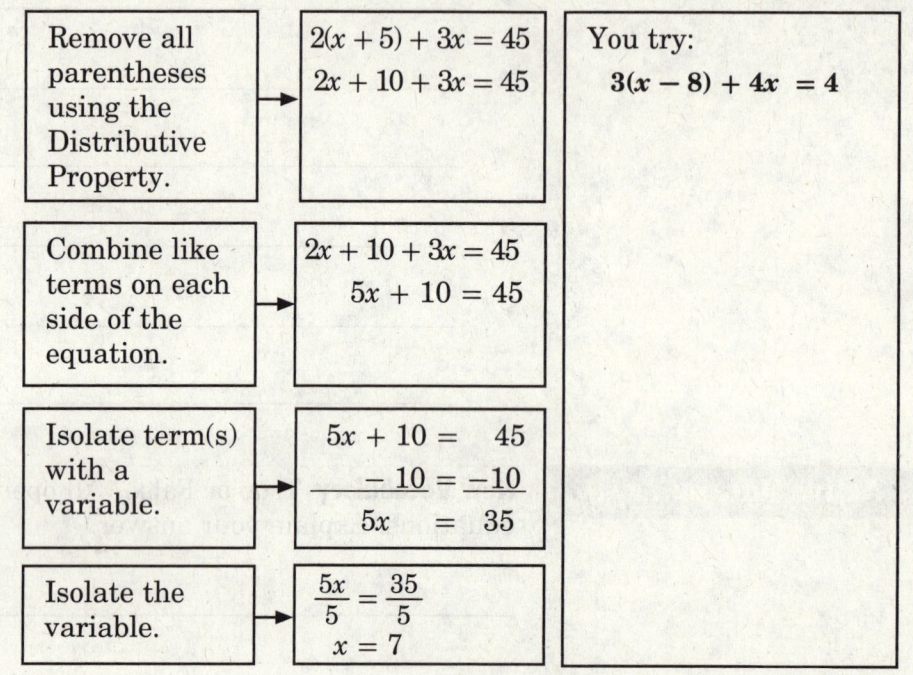
Solve Equations with Two Variables	**Write and solve an equation for the following situation.** *Mr. Ludwig wants to rent a post hole digger to build a deck. He pays a rate of $5 per hour and $12.50 for gas and insurance to rent the digger. what is the cost for a six-hour rental?* The cost of the _____ is a flat rate. The variable is the number of _____ h for which he rents the digger. _____ _____

Helping You Remember   Look up the word *solution* in a dictionary. What is one meaning that relates to the way you use the word in algebra?

_____

_____

Chapter 1   12   Glencoe Algebra 1

NAME _____ DATE _____ PERIOD _____

1-6 Relations

What You'll Learn

Skim the lesson. Write two things you already know about relations.

1. _____

_____

_____

2. _____

_____

_____

Active Vocabulary

New Vocabulary Label the elements of the diagram with the correct terms.

x-coordinate

x-axis

y-coordinate

y-axis

ordered pair

origin

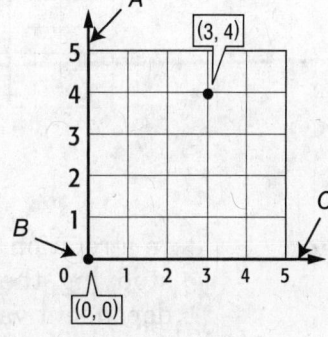

1. The numbers (3, 4) represent a(n) _____.

2. In the ordered pair (3, 4), 3 represents a(n) _____.

3. In the ordered pair (3, 4), 4 represents a(n) _____.

4. Arrow "A" is pointing to the _____.

5. Arrow "B" is pointing to the _____.

6. Arrow "C" is pointing to the _____.

Chapter 1   13   Glencoe Algebra 1

Lesson 1-6 (continued)

Main Idea	Details
Represent a Relation	Complete the table, graph, and mapping to represent the same relation shown below.

ordered pairs

(0, 1)
(3, 3)
(4, 2)

1. table

x	y
0	
3	
4	

2. graph

3. mapping

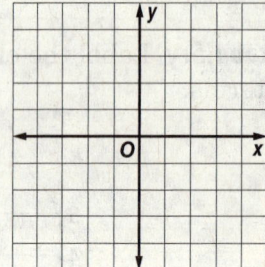

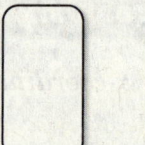

Graphs of a Relation   In a relation involving test grades, the more hours spent studying, the higher the grade. Identify the independent and dependent variables.

_____

_____

Helping You Remember   In the alphabet, *x* comes before *y*. Use this fact to describe a method for remembering how to write ordered pairs.

_____

_____

NAME _____ DATE _____ PERIOD _____

1-7 Functions

What You'll Learn Skim the lesson. Predict two things that you expect to learn based on the headings and the Key Concept box.

1. _____

_____

2. _____

_____

Active Vocabulary **Review Vocabulary** Define *reciprocals* in your own words. *(Lesson 1-3)*

_____

_____

New Vocabulary Fill in the blanks with the correct term or phrase.

function A graph that consists of points that are not connected is a _____.

discrete function A _____ is a function graphed with a line or a smooth curve.

continuous function A _____ is a relationship between input and output.

vertical line test A test used to determine whether or not a graph represents a function is known as the _____.

Vocabulary Link *Function* is a word that is used in everyday English. Find the definition of *function* using a dictionary. Explain how the English definition can help you remember how *function* is used in mathematics.

_____

_____

_____

Lesson 1-7 (continued)

Main Idea | Details

Identify Functions

Fill in each blank to tell how to determine if a relation is a function.

1.
x	y
1	-2
3	4
4	6
1	5

2.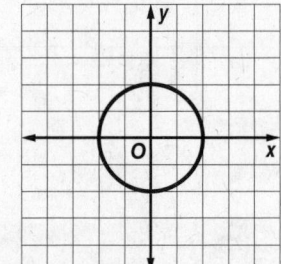

Look at a table to see if each member of the _____ corresponds to only _____ member of the _____.

Use the _____ test. If no vertical line can be drawn so that it intersects the graph more than _____, then it is a function.

Find Function Values

For $f(x) = 7x - 4$, find each value.

1. $f(3)$ _____   2. $f(-2)$ _____

3. $f(0)$ _____   4. $f(-3)$ _____

Helping You Remember

A student who was trying to help a friend remember how functions are different from relations that are not functions gave the following advice: *Just remember that functions are very strict and never give you a choice.* Explain how this might help you remember what a function is.

_____

_____

_____

_____

1-8 Interpreting Graphs of Functions

What You'll Learn

Scan the text under the *Now* heading. List two things you will learn about in this lesson.

1. _____

2. _____

Active Vocabulary

Review Vocabulary Write the definition next to each term. *(Lessons 1–6 and 1–7)*

domain ▶ _____

range ▶ _____

x-coordinate ▶ _____

y-coordinate ▶ _____

New Vocabulary Match each term with its definition.

x-intercept	the x-coordinate of a point where a graph crosses the x-axis
end behavior	the y-coordinate of a point where a graph crosses the y-axis
extrema	how the values of a function behave at the each end of the graph
y-intercept	when a function is divided by a vertical line into two halves that match exactly
line symmetry	the minimum or maximum values of a function
increasing/decreasing	the graph when it lies above the x-axis or below the x-axis
positive/negative	the graph as it goes up or as it goes down when viewed from left to right

Chapter 1 — Glencoe Algebra 1 — Lesson 1-8

Lesson 1-8 (continued)

Main Idea	Details
Intercepts and Symmetry	Referring to the graph of f(x) identify the function as linear or nonlinear. Then estimate and interpret the intercepts. 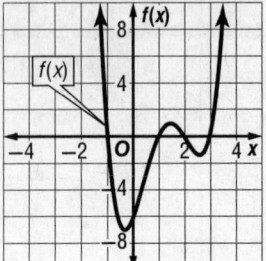

linear or nonlinear: _____

x-intercept(s): _____

y-intercept: _____

Extrema and End Behavior

Referring to the graph of f(x) estimate and interpret where the function is positive, negative, increasing, and decreasing, the x-coordinates of any relative extrema, and the end behavior of the graph.

positive: _____

negative: _____

increasing: _____

decreasing: _____

extrema: _____
_____

end behavior: _____
_____
_____

Helping You Remember A good way to remember something is to explain it to someone else. Suppose one of your classmates is having trouble understanding the difference between end behavior and extrema. How can you explain the different to him or her?

_____

_____

_____

_____

Chapter 1   18   Glencoe Algebra 1

Chapter 1: Expressions, Equations, and Functions

Tie It Together

Add details to each part of the graphic organizer.

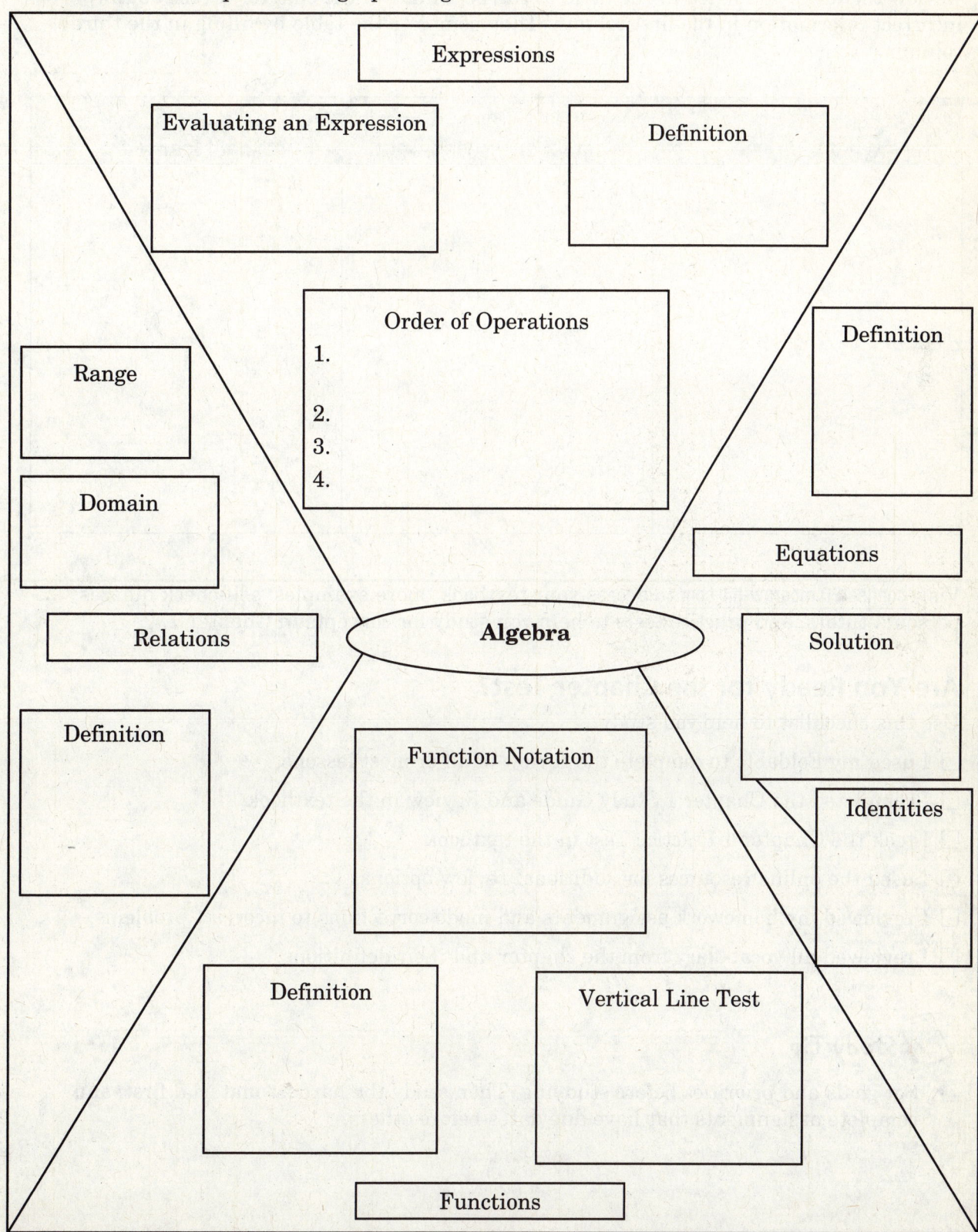

NAME _____ DATE _____ PERIOD _____

Expressions, Equations, and Functions

Before the Test

Review the ideas you listed in the table at the beginning of the chapter. Cross out any incorrect information in the first column. Then complete the table by filling in the third column.

K What I know...	W What I want to find out...	L What I learned...

Visit **connectED.mcgraw-hill.com** to access your textbook, more examples, self-check quizzes, personal tutors, and practice tests to help you study for concepts in Chapter 1.

Are You Ready for the Chapter Test?

Use this checklist to help you study.

☐ I used my Foldable to complete the review of all or most lessons.

☐ I completed the Chapter 1 Study Guide and Review in the textbook.

☐ I took the Chapter 1 Practice Test in the textbook.

☐ I used the online resources for additional review options.

☐ I reviewed my homework assignments and made corrections to incorrect problems.

☐ I reviewed all vocabulary from the chapter and their definitions.

- Set goals and priorities before studying. Then study the hardest material first, and complete assignments that have due dates before others.

NAME _____ DATE _____ PERIOD _____

Chapter 2 Linear Equations

Before You Read

Before you read the chapter, think about what you know about linear equations. List three things you already know about them in the first column. Then list three things you would like to learn about them in the second column.

K What I know...	W What I want to find out...

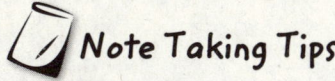 Construct the Foldable as directed at the beginning of this chapter.

Note Taking Tips

- **When you take notes, circle, underline, or star anything the teacher emphasizes.**
  When your teacher emphasizes a concept, it will usually appear on a test, so make an effort to include it in your notes.

- **Before going to class, look over your notes from the previous class, especially if the day's topic builds from the last one.**

NAME _____ DATE _____ PERIOD _____

CHAPTER 2 Linear Equations

Key Points

Scan the pages in the chapter and write at least one specific fact concerning each lesson. For example, in the lesson on ratios and proportions, one fact might be that the ratio of two measurements having different units of measure is called a rate. After completing the chapter, you can use this table to review for your chapter test.

Lesson	Fact
2-1 Writing Equations	
2-2 Solving One-Step Equations	
2-3 Solving Multi-Step Equations	
2-4 Solving Equations with the Variable on Each Side	
2-5 Solving Equations Involving Absolute Value	
2-6 Ratios and Proportions	
2-7 Percent of Change	
2-8 Literal Equations and Dimensional Analysis	
2-9 Weighted Averages	

NAME _____ DATE _____ PERIOD _____

2-1 Writing Equations

What You'll Learn

Skim the lesson. Write two things you already know about writing equations.

1. _____
   _____

2. _____
   _____

Active Vocabulary

Review Vocabulary Define *equation* in your own words. (*Lesson 1-5*)

_____

_____

New Vocabulary Define the term *formula* from this lesson.

_____

_____

Vocabulary Link *Formula* is a word that is used in everyday English. Find the definition of *formula* using a dictionary. Explain how its English definition can help you understand the meaning of *formula* in mathematics.

_____

_____

_____

_____

Chapter 2　　　　　23　　　　　Glencoe Algebra 1

Lesson 2-1 (continued)

Main Idea	Details
Write Verbal Expressions	Use a model to help translate the sentence below into an equation. *Six more than a number squared is 30 less than five times the number.* 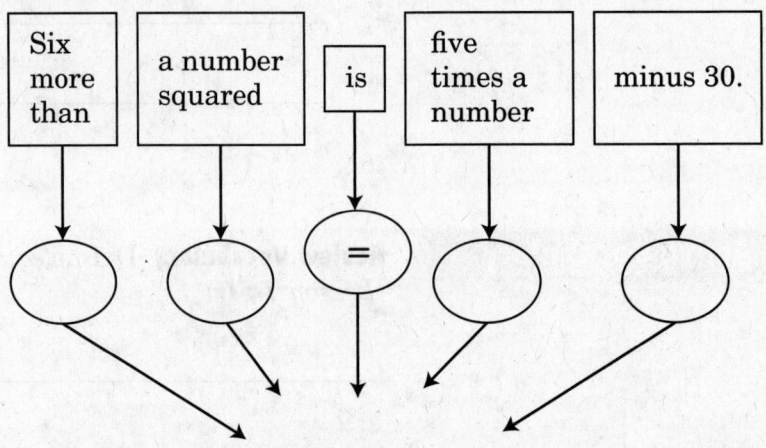
Write Sentences from Equations	Translate each equation into a sentence. 1. $7x + 2 = 30$ _____ _____ 2. $p^2 + 18 = 7 - k$ _____ _____

Helping You Remember If you cannot remember all the steps of the Four-Step Problem-Solving Plan, try to remember the first letters of the first word in each step. Write the associated words for each letter.

U _____ P _____ S _____ C _____

Chapter 2   24   Glencoe Algebra 1

NAME _____ DATE _____ PERIOD _____

2-2 Solving Equations

What You'll Learn

Skim the lesson. Predict two things that you expect to learn based on the headings and the Key Concept box.

1. _____

_____

2. _____

_____

_____

Active Vocabulary

Review Vocabulary Define *formula* in your own words. *(Lesson 2-1)*

_____

_____

New Vocabulary Fill in each blank with the correct term or phrase.

equivalent equations   To find the value of the variable that makes the equation true is to _____.

solve an equation   _____ have the same solution.

Vocabulary Link *Solution* is a word that is used in everyday English. Find the definition of *solution* using a dictionary. Explain how its English definition can help you understand its meaning in mathematics.

_____

_____

_____

_____

Chapter 2   25   Glencoe Algebra 1

Lesson 2-2 (continued)

Main Idea	Details
Solving Equations Using Addition or Subtraction	Adding the same quantity to two equal or "balanced" amounts, will yield scales that remain balanced.

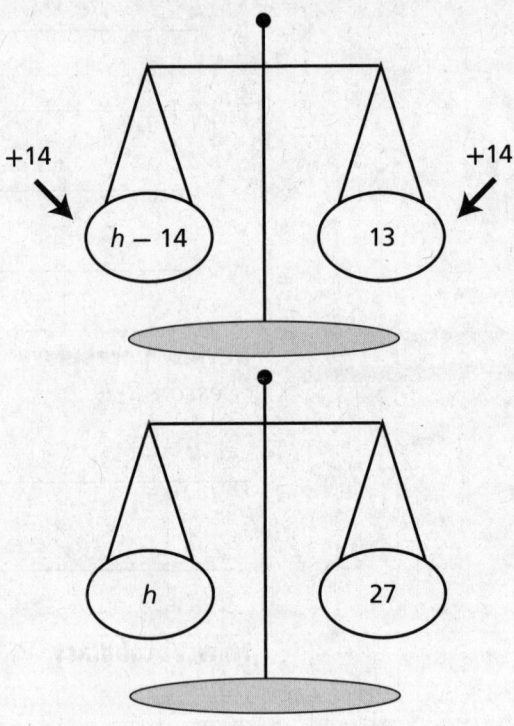

| **Solving Equations Using Multiplication or Division** | Solve by multiplying.

1. $\dfrac{g}{4} = 7$ 2. $\dfrac{m}{-5} = 3$

Solve by dividing.

3. $9y = 108$ 4. $5k = -115$ |

Helping You Remember

One way to remember something is to explain it to someone else. Write how you would explain to a classmate how to solve the equation $\dfrac{2}{3}x = 12$.

_____

_____

NAME _____ DATE _____ PERIOD _____

2-3 Solving Multi-Step Equations

What You'll Learn Scan the text under the *Now* heading. List two things you will learn about solving multi-step equations.

1. _____

   _____

2. _____

   _____

Active Vocabulary

Review Vocabulary Define *equivalent equations* in your own words. *(Lesson 2-2)*

_____

_____

New Vocabulary Write the correct term next to each definition.

_____ ▶ integers in counting order

_____ ▶ the study of numbers and the relationships between them

_____ ▶ an equation that requires more than one step to solve

Vocabulary Link *Consecutive* is a word that is used in everyday English. Find the definition of *consecutive* using a dictionary. Explain how its English definition can help you understand the meaning of *consecutive* in mathematics.

_____

_____

_____

NAME _____ DATE _____ PERIOD _____

Lesson 2-3 (continued)

Main Idea	Details
Solve Multi-Step Equations	**Solve the equation.**

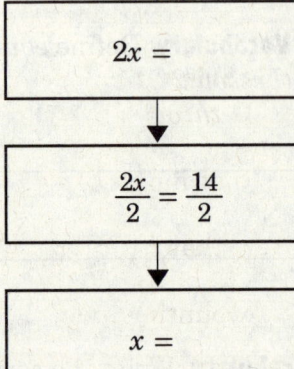

$2x + 3 = 17$	Original equation
$2x + 3 \quad\quad = 17$	Subtract from each side.
$2x =$	Simplify.
$\dfrac{2x}{2} = \dfrac{14}{2}$	Divide each side.
$x =$	Simplify.

Solve Consecutive Integer Problems

Write an equation for the following problem. Then solve the equation and answer the problem.

Find three consecutive even integers with a sum of 48.

Helping You Remember

Explain why working backward is a useful strategy for solving equations.

_____

_____

Chapter 2   28   Glencoe Algebra 1

NAME _____ DATE _____ PERIOD _____

2-4 Solving Equations with the Variable on Each Side

What You'll Learn

Skim the Examples in the lesson. Predict two things you think you will learn about solving equations with the variable on each side.

1. _____
   _____
   _____

2. _____
   _____
   _____

Active Vocabulary

Review Vocabulary Match each term with its definition. *(Lessons 2-1 through 2-3)*

formula	equations that have the same solution
solve an equation	an equation that requires more than one step to solve.
number theory	integers in counting order
equivalent equations	the study of numbers and the relationships between them
multi-step equations	a rule for the relationship between certain quantities
consecutive integers	finding the value of the variable that makes an equation true

New Vocabulary Define *identity* in your own words.

_____
_____

Vocabulary Link *Identity* is a word that is used in everyday English. Find the definition of *identity* using a dictionary. Explain how the English definition can help you remember how it is used in mathematics.

_____
_____
_____

NAME _____ DATE _____ PERIOD _____

Lesson 2-4 (continued)

Main Idea	Details

Variables on Each Side	Complete the *flow chart* to describe the steps in solving the equation.
	$7(2x - 3) = 12x - 5$
	
	┌─────────────────────────────────────┐
	│ Use the _____ Property. │
	│ _____ = 12x − 5 │
	└─────────────────────────────────────┘
	            ↓
	┌─────────────────────────────────────┐
	│ _____ 12x from each side and simplify. │
	│ ___ − 21 = ___ │
	└─────────────────────────────────────┘
	            ↓
	┌─────────────────────────────────────┐
	│ Add ___ to each side and simplify. │
	│ 2x = ___ │
	└─────────────────────────────────────┘
	            ↓
	┌─────────────────────────────────────┐
	│ _____ each side by 2. │
	│ x = ___ │
	└─────────────────────────────────────┘

Grouping Symbols	Solve the equation $6y + 4 = 3(2y - 10)$.

Helping You Remember   In addition to the examples in this section of Chapter 2, there will be other occurrences of *no solutions*, as well as *identities* where there are endless possibilities of solutions. What are the symbols for these?

NAME _____ DATE _____ PERIOD _____

2-5 Solving Equations Involving Absolute Value

What You'll Learn

Scan the text in the lesson. Write two facts you learned about solving equations involving absolute value as you scanned the text.

1. _____

_____

_____

_____

2. _____

_____

_____

_____

_____

Active Vocabulary

Review Vocabulary Label the elements of the diagram with the correct terms. *(Lesson 1-1)*

$$\underbrace{2\overset{B\downarrow}{y^3} + 5\overset{C\downarrow}{y} - 8}_{A\uparrow}$$

algebraic expression 1. The term 5y represents a(n) _____.

power 2. Arrow "A" is pointing to a(n) _____.

product 3. Arrow "B" is pointing to a(n) _____.

variable 4. Arrow "C" is pointing to a(n) _____.

Define *absolute value* in your own words.

_____

_____

_____

Chapter 2   31   Glencoe Algebra 1

Lesson 2-5 (continued)

Main Idea	Details
Absolute Value Expressions	Evaluate the following absolute value expression in the space provided. $\|f + 7\| - 11$ if $f = -9$
	Complete the organizer below. 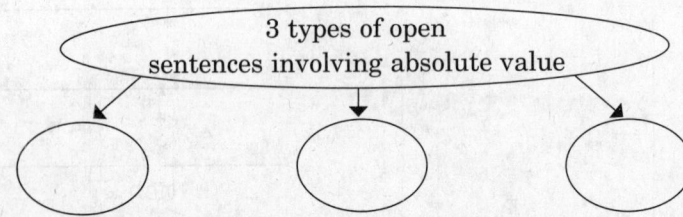
Absolute Value Equations	Write an absolute value equation that fits the solution graphed below. Then, write the solution set. 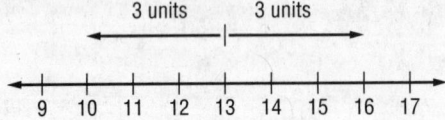

Helping You Remember What is one way you could check to see that your graph of an absolute value equation is correct?

_____

_____

_____

_____

Chapter 2   32   Glencoe Algebra 1

2-6 Ratios and Proportions

What You'll Learn

Scan the lesson. List two headings you would use to make an outline of this lesson.

1. _____

_____

2. _____

_____

Active Vocabulary

New Vocabulary Fill in each blank with the correct term or phrase.

unit — In the proportion 2:5 = 6:15, the numbers 5 and 6 are known as the _____.

ratio — A(n) _____ is an equation stating that two ratios are equal.

means — A(n) _____ is a rate used when making a model of something that is too large or too small to be convenient at actual size.

rate — The comparison of two numbers by division is known as a(n) _____.

model — A _____ rate tells how many of one item is being compared to one of another item.

extremes — In the proportion 1:15 = 3:45, the numbers 1 and 45 are known as the _____.

proportion — The ratio of two measurements having different units of measure is called a(n) _____.

scale — A scale _____ is a three-dimensional reproduction of an item that has been reduced in size proportionally.

Chapter 2   33   Glencoe Algebra 1

Lesson 2-6 (continued)

Main Idea	Details
Ratios and Proportions	Use cross products to determine whether the pair of ratios forms a proportion. $\frac{3}{4}, \frac{4.2}{6}$ _____ _____
Solve Proportions	Use the graphic organizer below to help solve the rate of growth proportion that follows. A women's exercise franchise opened 336 gyms during the past 3 years. If their growth rate remains constant, how many exercise gyms will they have opened after 5 years? Let g represent the number of gyms. $\boxed{\frac{\text{number of gyms}}{\text{number of years}}} \longrightarrow \boxed{\frac{\text{gyms}}{3 \text{ years}}} = \boxed{\frac{\text{gyms}}{5 \text{ years}}}$ _____ _____ _____

Helping You Remember What is one way you could check to see that your graph of an absolute value equation is correct?

_____

_____

NAME _____ DATE _____ PERIOD _____

2-7 Percent of Change

What You'll Learn

Scan the lesson. List two headings you would use to make an outline of this lesson.

1. _____

_____

2. _____

_____

Active Vocabulary

Review Vocabulary Match each term with its definition. *(Lesson 2-6)*

proportion   a ratio of two measurements having different units of measure

ratio   an equation stating that two ratios are equal

rate   a comparison of two numbers by division

New Vocabulary Fill in each blank with the correct term or phrase.

_____ is the ratio of the change in an amount to the original amount expressed as a percent.

When the new number is less than the original number, the percent of change is a percent of _____.

When the new number is greater than the original number, the percent of change is a percent of _____.

Vocabulary Link *Change* is a word that is used in everyday English. Find the definition of *change* using a dictionary. Explain how the English definition can help you remember how *change* is used in mathematics.

_____

_____

_____

Chapter 2   35   Glencoe Algebra 1

Lesson 2-7 (continued)

Main Idea	Details
Percent of Change	Use the graphic organizer to help you find the percent of change given an original amount of 30, and a final amount of 45.

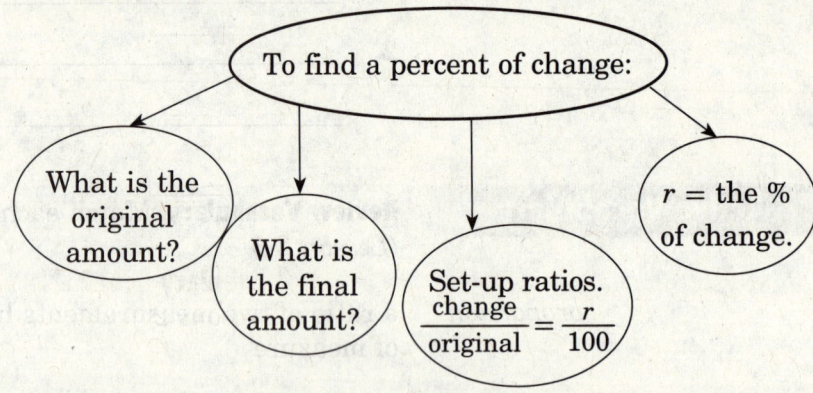

The amount of change = _____ − _____ = _____.

$\frac{15}{45} = \frac{r}{100}$ ... solve for r and get $r =$ _____

Therefore, the percent of change is a _____ % _____ (increase or decrease).

Solve Problems — Tess purchased a dress that originally cost $110. The day she made the purchase it was on sale for 20% off. What was the sale price of her dress?

_____

Helping You Remember

If you remember only two things about the ratio used for finding the percent of change, what should they be?

_____
_____
_____

2-8 Literal Equations and Dimensional Analysis

What You'll Learn

Scan the text under the *Now* heading. List two things you will learn about in this lesson.

1. _____

   _____

2. _____

   _____

Active Vocabulary

Review Vocabulary Fill in each blank with the correct term or phrase. *(Lessons 2-1, 2-4, and 2-7)*

_____ are equations that are true for all values of the variables.

The ratio of the change in an amount to the original amount expressed as a percent is known as the _____.

A(n) _____ is a rule for the relationship between certain quantities.

New Vocabulary Match each term with its definition.

dimensional analysis   a formula or equation that involves several variables

literal equation   the process of carrying units throughout a computation

unit analysis   another term for dimensional analysis

Vocabulary Link *Literal* is a word that is used in everyday English. Find the definition of *literal* using a dictionary. Explain how the English definition can help you remember how *literal* is used in mathematics.

_____

_____

_____

_____

NAME _____ DATE _____ PERIOD _____

Lesson 2-8 (continued)

Main Idea	Details
Solve for a Specific Variable	Fill in the missing pieces of the graphic organizer below.

Solve for p.

Isolate the terms with that variable onto one side of the equation.
→
$4p - 3q = pr + 9$
$ + 3q = + 3q$
$4p = \underline{}$
$-pr = -pr$
$\underline{}$

_____ Property → $p(4 - r) = 3q + 9$

_____ each side by → $\dfrac{p(4 - r)}{4 - r} = \dfrac{3q + 9}{4 - r}$

Simplify. → $p = \dfrac{3q + 9}{4 - r}$, $r \neq$ _____

Use Formulas

The formula for the area of a rectangle is $A = \ell w$, where ℓ is its length, and w is the width. Find the length of a rectangular garden that has an area of 5400 square feet and a width of 90 feet.

_____

Helping You Remember

When you give the dimensions of a rectangle, you have to tell how many units long it is and how many units wide it is. How can this help you remember what dimensional analysis involves?

_____

_____

_____

Chapter 2   Glencoe Algebra 1

NAME _____ DATE _____ PERIOD _____

2-9 Weighted Averages

What You'll Learn

Scan the text under the *Now* heading. List two things you will learn about in the lesson.

1. _____

2. _____

Active Vocabulary

New Vocabulary Match each term with its definition.

rate problem — When referring to a set of data, it is the sum of the product of the number of units and the value per unit divided by the sum of the number of units.

uniform motion problem — problems in which two or more parts are combined into a whole

weighted average — problems in which an object moves at a certain speed or rate

mixture problem — a statement that requires a solution, usually by means of a mathematical operation

Vocabulary Link *Problem* is a word that is used in everyday English. Find the definition of *problem* using a dictionary. Explain how the English definition can help you remember how *problem* is used in mathematics.

_____

_____

_____

_____

Chapter 2   39   Glencoe Algebra 1

Lesson 2-9 (continued)

Main Idea	Details
Weighted Averages	In addition to the slugging average that was in the textbook, write another example of when it might be necessary to calculate a weighted average. _____ _____ _____ _____
Uniform Motion Problems	Use the table provided to aid in solving the following rate problem. Two cyclists begin traveling from opposite ends of a 15-kilometer bike path towards each other. One of the cyclists is traveling 20 kilometers per hour, and the other cyclist is traveling 25 kilometers per hour. How much time will it take for them to meet each other? 		r	t	$d = rt$	 |---|---|---|---| 	first cyclist				 	second cyclist				 Now write and solve an equation. ____ + ____ = 15 ____ $t = 15$ $t =$ ____ or ____ hour or ____ minutes

Helping You Remember Making a table can be helpful in solving mixture problems. In your own words, explain how you use a table to solve mixture problems.

_____

_____

_____

Chapter 2 Linear Equations

Tie It Together

Provide the indicated details in each graphic organizer.

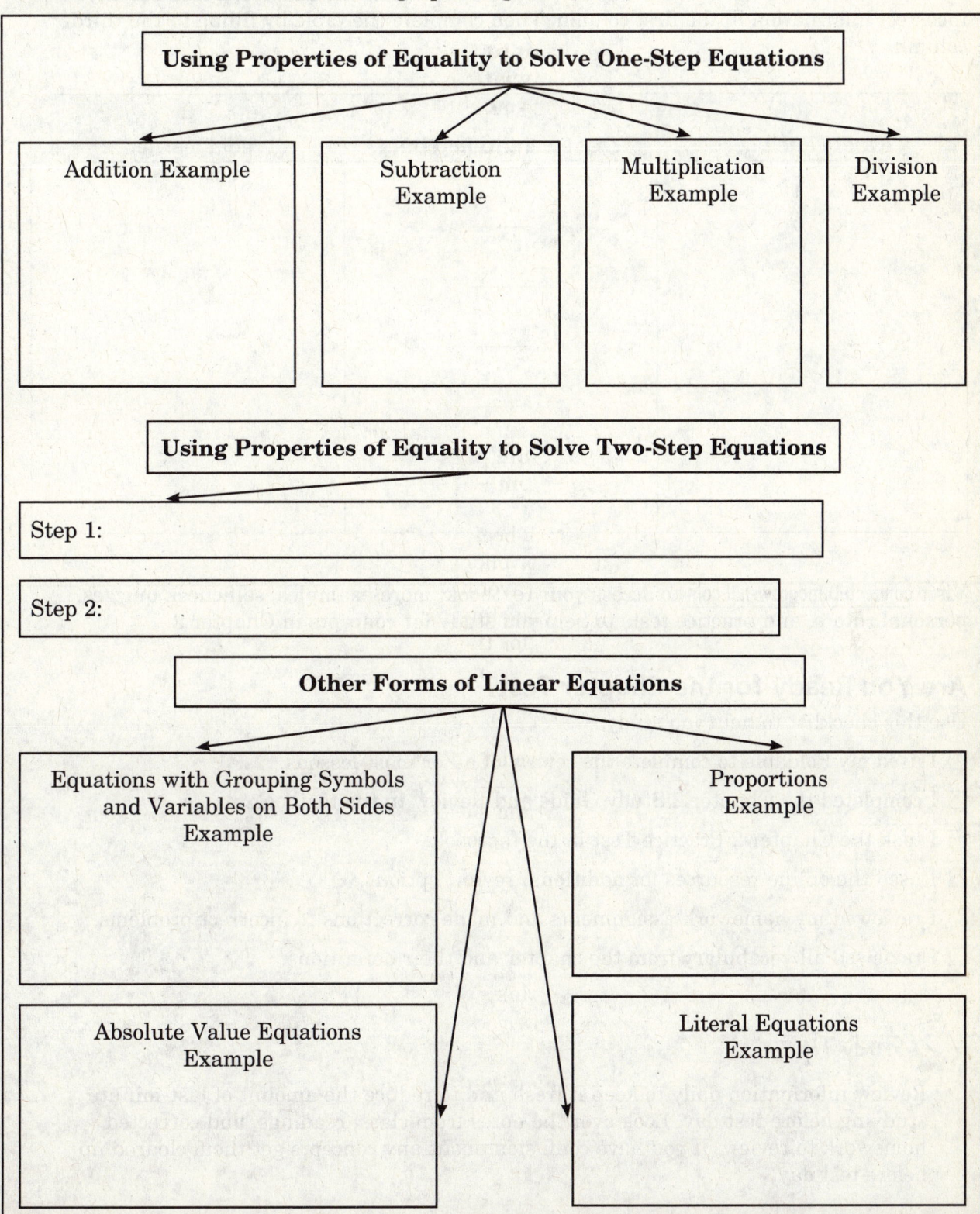

NAME _____ DATE _____ PERIOD _____

CHAPTER 2 Linear Equations

Before the Test

Review the ideas you listed in the table at the beginning of the chapter. Cross out any incorrect information in the first column. Then complete the table by filling in the third column.

K What I know...	W What I want to find out...	L What I learned...

Visit connectED.mcgraw-hill.com to access your textbook, more examples, self-check quizzes, personal tutors, and practice tests to help you study for concepts in Chapter 2.

Are You Ready for the Chapter Test?

Use this checklist to help you study.

☐ I used my Foldable to complete the review of all or most lessons.

☐ I completed the Chapter 2 Study Guide and Review in the textbook.

☐ I took the Chapter 2 Practice Test in the textbook.

☐ I used the online resources for additional review options.

☐ I reviewed my homework assignments and made corrections to incorrect problems.

☐ I reviewed all vocabulary from the chapter and their definitions.

Study Tip

- Review information daily to keep it fresh and to reduce the amount of last-minute studying before test day. Look over the notes from class, readings, and corrected homework to review. If you have confusion about any concepts get them cleared up before test day.

Chapter 3: Linear Functions

Before You Read

Before you read the chapter, respond to these statements.
1. Write an **A** if you agree with the statement.
2. Write a **D** if you disagree with the statement.

Before You Read	Linear Functions
	• The graph of a linear equation is a straight line.
	• A family of graphs is different equations that represent the same line.
	• Slope and rate of change are the same thing.
	• Slope is the change in x over the change in y.
	• The graph of a nonproportional relationship will not be a straight line.

 Construct the Foldable as directed at the beginning of this chapter.

 Note Taking Tips

- **When you take notes, write down the math problem and each step in the solution using math symbols.**
  Next to each step, write down, in your own words, exactly what you are doing.
- **It is helpful to read through your notes before beginning your homework.**
  Look over any page referenced material.

NAME _____ DATE _____ PERIOD _____

CHAPTER 3
Linear Functions

Key Points

Scan the pages in the chapter and write at least one specific fact concerning each lesson. For example, in the lesson on solving linear equations by graphing, one fact might be that the root of an equation is any value that makes the equation true or the solution. After completing the chapter, you can use this table to review for your chapter test.

Lesson	Fact
3-1 Graphing Linear Equations	
3-2 Solving Linear Equations by Graphing	
3-3 Rate of Change and Slope	
3-4 Direct Variation	
3-5 Arithmetic Sequences as Linear Functions	
3-6 Proportional and Nonproportional Relationships	

3-1 Graphing Linear Equations

What You'll Learn

Scan the lesson. List two headings you would use to make an outline of this lesson.

1. _____

2. _____

Active Vocabulary

New Vocabulary Match the term with its definition by drawing a line to connect the two.

linear equation	the x-coordinate of the point at which the graph of an equation crosses the x-axis
standard form	a number
x-intercept	an equation which forms a line when it is graphed
constant	the y-coordinate of the point at which the graph of an equation crosses the y-axis
y-intercept	a linear equation written in the form $Ax + By = C$

Vocabulary Link Determine whether each of the following is a linear equation. Using a graphing calculator, sketch a graph of each equation.

$y = 3x - 4$	$y = 3x^2 - 4$	$y = 0x - 4$
Linear? Yes No	Linear? Yes No	Linear? Yes No

Lesson 3-1 (continued)

Main Idea	Details
Identify Linear Equations and Intercepts	Write a word problem that could be represented by the table of values. Label the independent variable and the dependent variable in the table. Graph the table of values, labeling the axes appropriately.

x	0	1	2	3	4
y	100	75	50	25	0

Word Problem

Graph Linear Equations — Describe the similarities and differences in finding the x-intercept of a line and finding the y-intercept of a line.

Similarities	Differences

Chapter 3 — Glencoe Algebra 1

3-2 Solving Linear Equations by Graphing

What You'll Learn

Scan the text in the lesson. Write two facts you learned about solving linear equations by graphing as you scanned the text.

1. _____

2. _____

Active Vocabulary

Review Vocabulary Solve each equation for x. Label each as being *consistent*, *inconsistent*, or an *identity*. *(Lesson 2-3)*

$3x + 6 = 4x - 8$	$3x + 9 = 3x - 8$	$3x + 7 = 4x + 7 - x$

New Vocabulary Write the definition next to each term.

linear function ▶ _____

parent function ▶ _____

family of graphs ▶ _____

root ▶ _____

zeros ▶ _____

Chapter 3 — 47 — Glencoe Algebra 1

Lesson 3-2 (continued)

Main Idea	Details
Solve by Graphing	Complete the diagram to show the relationship between the words *root*, *solution*, *zero*, and *x-intercept*.

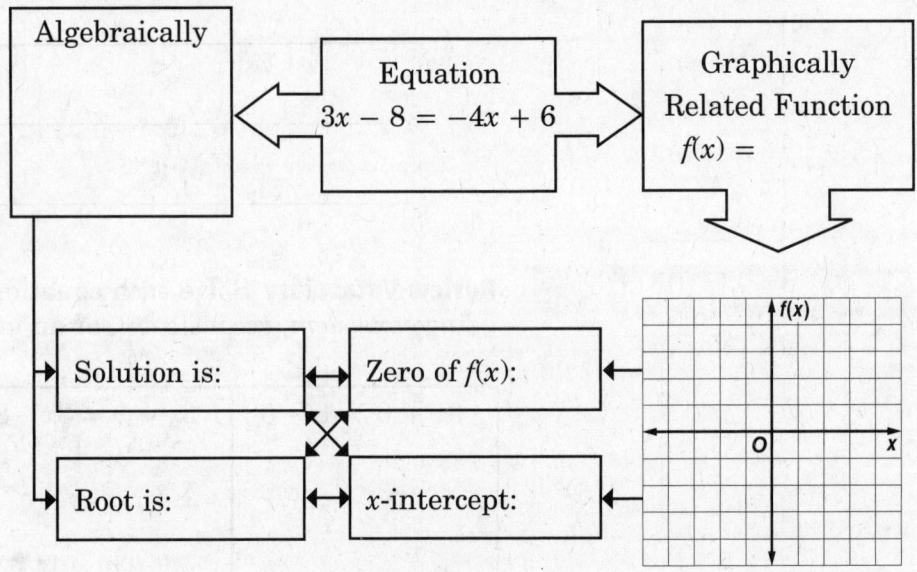

Estimate Solutions by Graphing	Write a function for the situation described below. Describe how to find the zero of this function. Determine what the zero of this function represents.

The salt reserve for a city's road crew was at 17 tons prior to the beginning of winter. Each time the roads are treated, the reserves are depleted by 3.25 tons of salt.

$f(x) = $ _____

Algebraically	Graphically	What does the zero represent?

Chapter 3   48   Glencoe Algebra 1

NAME _____ DATE _____ PERIOD _____

3-3 Rate of Change and Slope

What You'll Learn

Skim the lesson. Write two things you already know about rate of change and slope.

1. _____

   _____

2. _____

   _____

Active Vocabulary

Review Vocabulary Write the definition of the word *ratio* and list the three ways that a ratio can be expressed. By scanning ahead, what is a ratio used to represent in this lesson?

_____

_____

_____

_____

_____

New Vocabulary Write the definition next to each term.

rate of change ▶ _____

_____

_____

slope ▶ _____

_____

Chapter 3 49 Glencoe Algebra 1

Lesson 3-3 (continued)

Main Idea	Details
Rate of Change	Complete the table of values so that Table A has a *constant rate of change* of 20 gallons per hour and Table B has a *constant rate of change* of –15.5 inches per minute.

Table A

Hour	Gallons
1:00 P.M.	
4:00 P.M.	
6:00 P.M.	1250
10:00 P.M.	

Table B

Minutes	Inches
6	259.25
	228.25
12	
	42.25

Find Slope

Use each of the indicated methods to calculate the slope of the line described.

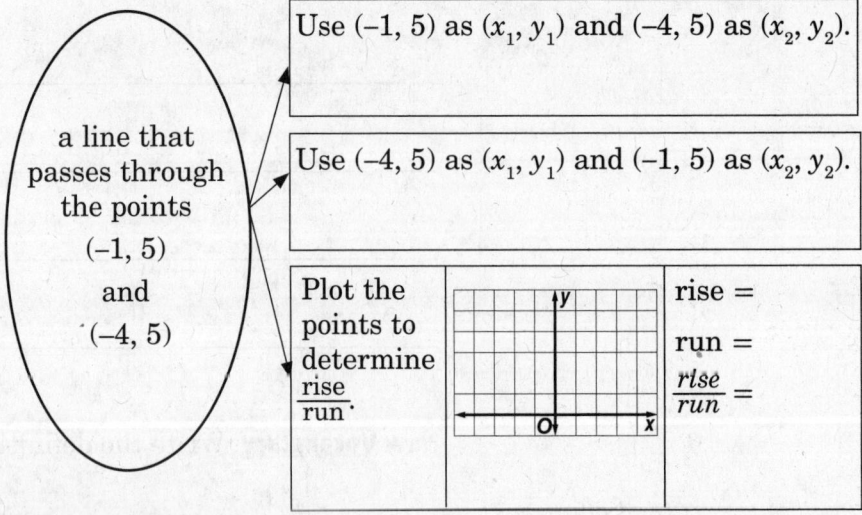

a line that passes through the points (–1, 5) and (–4, 5)

- Use (–1, 5) as (x_1, y_1) and (–4, 5) as (x_2, y_2).
- Use (–4, 5) as (x_1, y_1) and (–1, 5) as (x_2, y_2).
- Plot the points to determine $\frac{rise}{run}$.

rise =
run =
$\frac{rise}{run}$ =

Did you get the same slope all three times?

Helping You Remember The word *rise* is associated with going up. Sometimes going from one point to another on a graph does not involve a rise and a run but a fall and a run. Describe how you could select points so that it is always a rise from the first point to a second point.

3-4 Direct Variation

What You'll Learn

Skim the Examples in the lesson. Predict two things you think you will learn about direct variation.

1. _____

_____

2. _____

_____

Active Vocabulary

Review Vocabulary Write another possible point on each of the lines described. Use the slope formula to justify your answer. *(Lesson 3-3)*

1. passes through (5, 8) with negative slope

2. passes through (5, 8) with positive slope

3. passes through (5, 8) with zero slope

4. passes through (5, 8) with no slope

5. passes through (5, 8) with slope of 2

New Vocabulary Label the equation with the correct terms.

direct variation

$$y = kx$$

constant of variation

Chapter 3   51   Glencoe Algebra 1

Lesson 3-4 (continued)

Main Idea	Details
Direct Variation Equations	Complete the diagram by writing one characteristic of direct variation in each box. 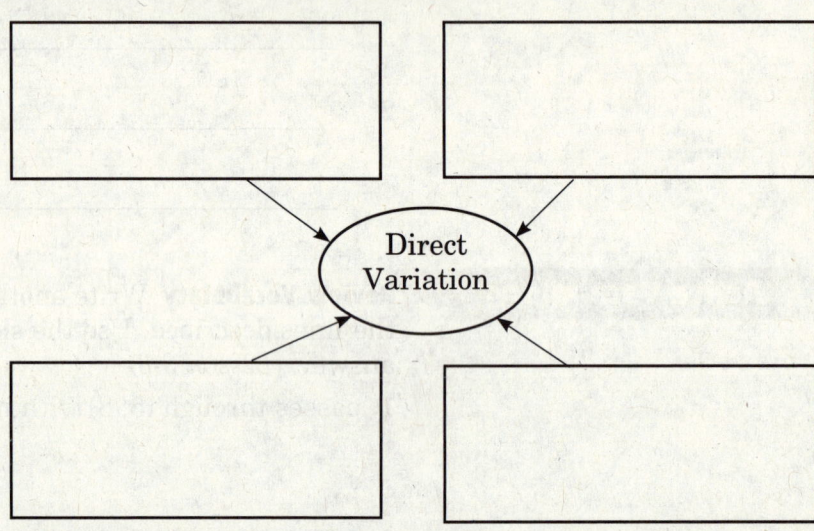
Direct Variation Problems	Write a direct variation equation for the situation described below. Determine Amanda's pay for 12 hours. *Amanda's paycheck varies directly as the number of hours that she works. If Amanda works 4 hours, her paycheck is $35.* 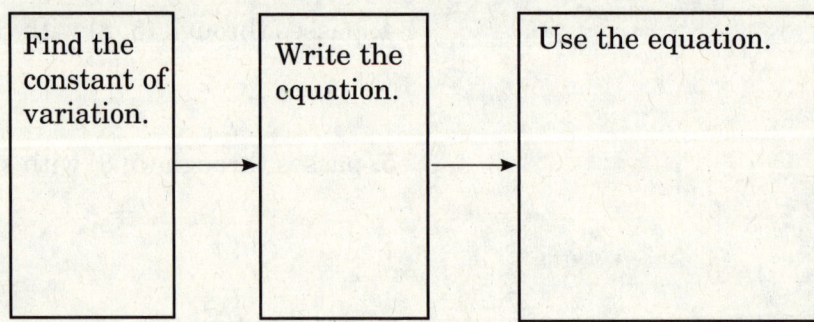

Helping You Remember Look up the word *constant* in a dictionary. How does this definition relate to the term *constant of variation*?

_____

_____

3-5 Arithmetic Sequences as Linear Functions

What You'll Learn

Skim the lesson. Predict two things that you expect to learn based on the headings and the Key Concept box.

1. _____

2. _____

Active Vocabulary

Review Vocabulary Evaluate $f(x) = 4x + 2$ and $g(x) = -3x + 7$ for $x = -1, 0, 1, 2, 3$. *(Lesson 3-2)*

x	−1	0	1	2	3
f(x)					
g(x)					

Describe the pattern you see in $f(x)$.

_____

Describe the pattern you see in $g(x)$.

_____

Describe the graph of the ordered pairs $(x, f(x))$.

_____

Describe the graph of the ordered pairs $(x, g(x))$.

_____

New Vocabulary Write the correct term beside each definition.

_____ ▶ the numbers in a sequence

_____ ▶ a sequence in which the difference in successive terms is constant

_____ ▶ a set of numbers in a specific order

_____ ▶ the difference between the terms in an arithmetic sequence

Chapter 3   53   Glencoe Algebra 1

Lesson 3-5 *(continued)*

Main Idea	Details
Recognize Arithmetic Sequences	Complete each question below. 1. Determine whether the sequence 3, −7, −14, −24, −31, −41 is an arithmetic sequence. Justify your answer. _____ _____ 2. Determine the next four terms of the arithmetic sequence −17, −12, −7, −2, 3, ... _____ 3. Write an equation for the nth term of the arithmetic sequence 14, 10, 6, 2, −2, ... _____
Arithmetic Sequences and Functions	Follow the steps below to write a function to represent the arithmetic sequence described. *Anya is collecting cans to turn into the recycling center. The arithmetic sequence $0.02, $0.04, $0.06, $0.08, ... represents the amount of money she earns for turning in the cans.* **Use the function to determine her earnings for turning in 100 cans.**

Determine the common difference.	→	

Substitute into the nth term formula. $a_n = a_1 + (n-1)d$	→	

Evaluate the function.	→	

Chapter 3   54   Glencoe Algebra 1

3-6 Proportional and Nonproportional Relationships

What You'll Learn

Scan the text under the *Now* heading. List two things you will learn about in the lesson.

1. _____
   _____
   _____

2. _____
   _____
   _____

Active Vocabulary

New Vocabulary Fill in the blanks with the correct terms or phrases.

inductive reasoning ▶ It is the process of using a _____ to make a general _____. When a _____ pattern is found, a linear equation can be written. The relationship is _____ if the linear equation is of the form $y = kx$.

Vocabulary Link Explain how the use of the word *proportional* in geometry can help you remember its use in this lesson.

_____

_____

_____

_____

NAME _____ DATE _____ PERIOD _____

Lesson 3-6 (continued)

Main Idea	Details
Proportional Relationships	Fill in the left boxes with details to describe how to determine whether a given relationship is proportional. Complete the example shown in the right boxes.

x	4	5	6	7	8
y	−12	−15	−18	−21	−24

Is the relationship linear?	Is the relationship linear?

↓ ↓

Does it pass through (0, 0)?	Does it pass through (0, 0)?

↓ ↓

Write an equation and check.	Write an equation and check.

Nonproportional Relationships

Describe how proportional and nonproportional relationships are similar. Describe how they are different.

_____

_____

_____

_____

Chapter 3   56   Glencoe Algebra 1

NAME _____ DATE _____ PERIOD _____

CHAPTER 3 Linear Functions

Tie It Together

Provide details in each graphic organizer.

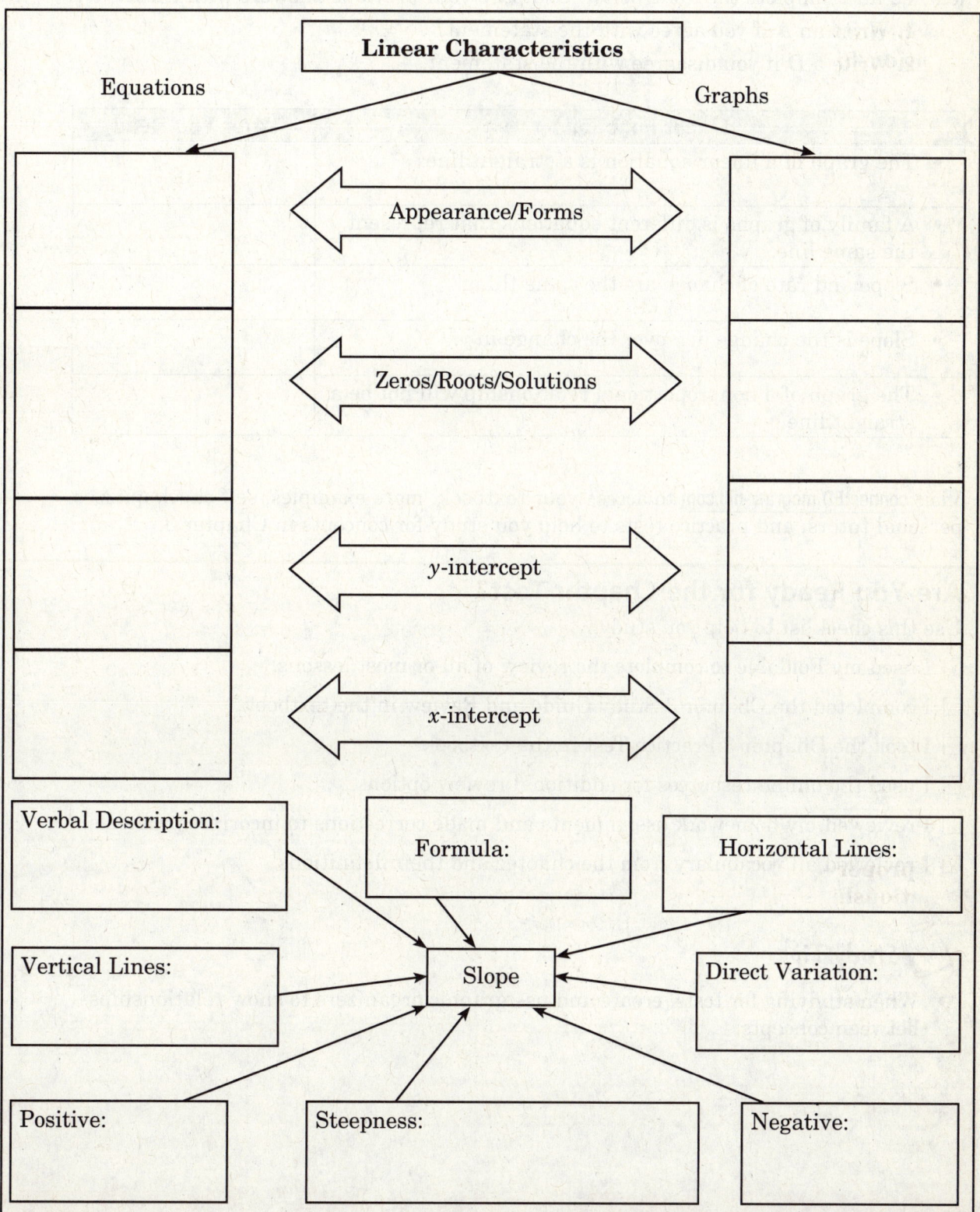

Chapter 3 — 57 — Glencoe Algebra 1

Chapter 3 Linear Functions

Before the Test

Now that you have read and worked through the chapter, think about what you have learned and complete the table below. Compare your previous answers with these.

1. Write an **A** if you agree with the statement.
2. Write a **D** if you disagree with the statement.

Linear Functions	After You Read
• The graph of a linear equation is a straight line.	
• A family of graphs is different equations that represent the same line.	
• Slope and rate of change are the same thing.	
• Slope is the change in x over the change in y.	
• The graph of a nonproportional relationship will not be a straight line.	

Visit **connectED.mcgraw-hill.com** to access your textbook, more examples, self-check quizzes, personal tutors, and practice tests to help you study for concepts in Chapter 3.

Are You Ready for the Chapter Test?

Use this checklist to help you study.

☐ I used my Foldable to complete the review of all or most lessons.

☐ I completed the Chapter 3 Study Guide and Review in the textbook.

☐ I took the Chapter 3 Practice Test in the textbook.

☐ I used the online resources for additional review options.

☐ I reviewed my homework assignments and made corrections to incorrect problems.

☐ I reviewed all vocabulary from the chapter and their definitions.

Study Tip

- When studying for tests, create and use graphic organizers to show relationships between concepts.

NAME _____ DATE _____ PERIOD _____

CHAPTER 4
Equations of Linear Functions

Before You Read

Before you read the chapter, think about what you know about equations of linear functions. List three things you already know about them in the first column. Then list three things you would like to learn about them in the second column.

K What I know...	W What I want to find out...

 Construct the Foldable as directed at the beginning of this chapter.

Note Taking Tips

- **As soon as possible, go over your notes.**
  Clarify any ideas that were not complete.

- **If you find it difficult to write and pay attention at the same time, write down key words only.**
  Then go back and complete your notes.

Chapter 4   59   *Glencoe Algebra 1*

NAME _____ DATE _____ PERIOD _____

Chapter 4: Equations of Linear Functions

Key Points

Scan the pages in the chapter and write at least one specific fact concerning each lesson. For example, in the lesson on scatter plots and lines of fit, one fact might be that scatter plots can show whether there is a trend in a set of data. After completing the chapter, you can use this table to review for your chapter test.

Lesson	Fact
4-1 Graphing Equations in Slope-Intercept Form	
4-2 Writing Equations in Slope-Intercept Form	
4-3 Writing Equations in Point-Slope Form	
4-4 Parallel and Perpendicular Lines	
4-5 Scatter Plots and Lines of Fit	
4-6 Regression and Median-Fit Lines	
4-7 Inverse Linear Functions	

NAME _____ DATE _____ PERIOD _____

4-1 Graphing Equations in Slope-Intercept Form

What You'll Learn Skim the lesson. Predict two things that you expect to learn based on the headings and the Key Concept box.

1. _____

2. _____

Active Vocabulary **Review Vocabulary** Identify the slope and *y*-intercept of lines A, B, C, and D. *(Lessons 3-1 and 3-3)*

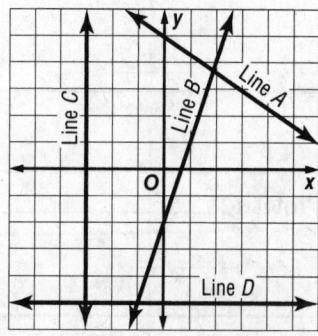

Line	Slope	y-intercept
A		
B		
C		
D		

New Vocabulary Label the diagram using the terms at the left.

slope-intercept form

y-intercept

slope

independent variable

dependent variable

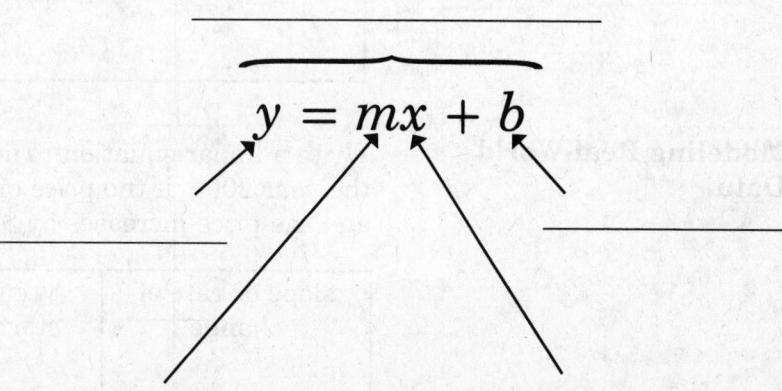

Chapter 4 61 Glencoe Algebra 1

NAME _____ DATE _____ PERIOD _____

Lesson 4-1 *(continued)*

Main Idea	Details
Slope-Intercept Form	Complete each step in the chart below. Add details to each step for clarification.

Write the equation in _____ form, if needed.

⬇

Identify the _____ and the _____.

⬇

Plot the _____ on a coordinate plane.

⬇

Plot another _____ using the _____.

Modeling Real-World Data

Write a linear equation to determine the price of gas after the year 2008, if the price of gas in 2008 is $3.16 per gallon and the price increases by $0.55 per year.

slope or rate of change	y-intercept or starting value	linear equation $y = mx + b$

Chapter 4   62   Glencoe Algebra 1

4-2 Writing Equations in Slope-Intercept Form

What You'll Learn

Skim the lesson. Write two things you already know about writing equations in slope-intercept form.

1. _____

_____

2. _____

_____

Active Vocabulary

Review Vocabulary Rewrite each equation in *slope-intercept form*. Circle the slope and underline the *y*-intercept. *(Lesson 4-1)*

$2y + 5x = -8$	$3y + 5x = 5x + 7$	$y + 5x = 4$

▶ linear extrapolation

New Vocabulary Fill in each blank with the correct terms.

a process in which you use a _____ equation to make _____ about a value that is outside the range of a given set of _____

Vocabulary Link Look up the word *extrapolate* in the dictionary. Write the non-mathematical definition of the word, a synonym for the word, and then use the word *extrapolate* in a non-mathematical sentence.

_____

_____

_____

_____

NAME _____ DATE _____ PERIOD _____

Lesson 4-2 (continued)

Main Idea	Details
Write an Equation Given the Slope and a Point	Fill in the diagram to write the equation of the line in *slope-intercept form*. Write an equation of the line that passes through (−2, 4) and has a slope of 2. 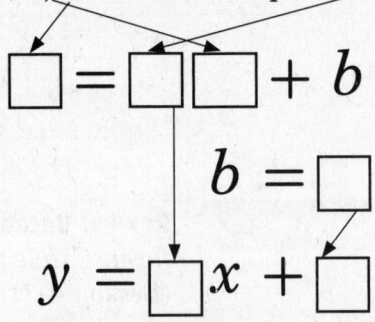 $\square = \square\square + b$ $b = \square$ $y = \square x + \square$
Write an Equation Given Two Points	Write an equation of the line that passes through (2, 4) and (−7, 5). Find the slope. → Find the *y*-intercept. → Write an equation.

Helping You Remember In your own words, explain how you would answer a question that asks you to write the slope-intercept form of an equation.

_____
_____
_____

4-3 Writing Equations in Point-Slope Form

What You'll Learn

Scan the text in the lesson. Write two facts you learned about equations in point-slope form as you scanned the text.

1. _____
   _____
   _____
   _____

2. _____
   _____
   _____
   _____

Active Vocabulary

New Vocabulary Label the diagram using the terms at the left.

dependent variable

slope

independent variable

x-coordinate of point on the line

y-coordinate of point on the line

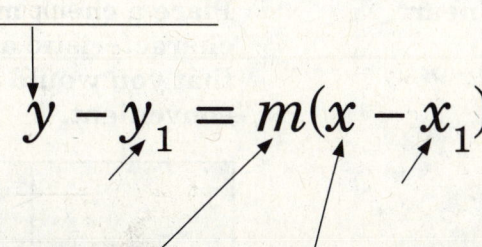

Vocabulary Link Write the point-slope formula and the slope formula below. Explain how the two formulas are related.

slope formula	point-slope form
How are they related?	

Chapter 4   Glencoe Algebra 1

Lesson 4-3

Lesson 4-3 (continued)

Main Idea	Details
Point-Slope Form	Write the equation of the line in *slope-intercept form* that passes through (−4, 5) and (6, −5) using the two different methods. Which method do you prefer? Explain. 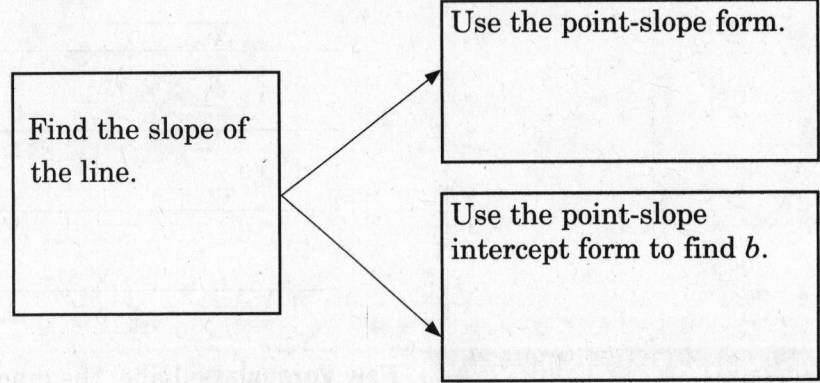
Forms of Linear Equations	Place a check mark in each box in which the specified characteristic applies. Describe the graphing method that you would use for each form identified as being convenient.

Form	Slope is easily identifiable.	The y-intercept is easily identifiable.	convenient form for graphing
point-slope form			
slope-intercept form			
standard form			

4-4 Parallel and Perpendicular Lines

What You'll Learn

Skim the Examples in the lesson. Predict two things you think you will learn about parallel and perpendicular lines.

1. _____

_____

_____

2. _____

_____

_____

_____

Active Vocabulary

Review Vocabulary Write the slope formula, and then write a verbal description of how to use the slope formula. *(Lesson 3-3)*

_____

_____

_____

_____

_____

New Vocabulary Write the correct term beside each definition.

_____ ▶ lines in the same plane that never intersect and have the same slope

_____ ▶ lines that intersect at right angles and have slopes that are opposite reciprocals

NAME _____ DATE _____ PERIOD _____

Lesson 4-4 (continued)

Main Idea	Details
Parallel Lines	Write an equation for each line described in *slope-intercept* form. 1. *x*-intercept of 3; *y*-intercept of -1 2. parallel to the line in Exercise 1 3. intersects the line in Exercise 1 at the *y*-intercept
Perpendicular Lines	Given two equations in standard form, determine whether the lines are parallel, perpendicular, or neither. Write each equation in slope-intercept form. $3x - 4y = 12$ $6y = 8x - 12$ $y = \underline{} x - 3$ $y = \underline{} x - 2$ 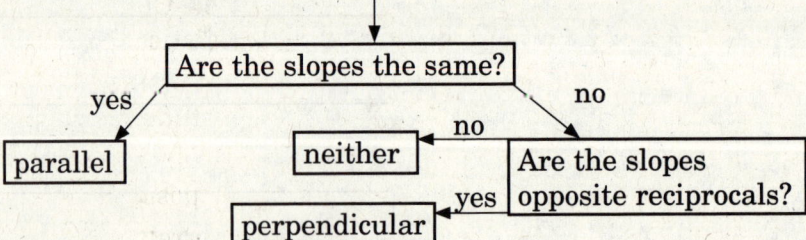

Helping You Remember   Explain to another person how you would use the *y*-intercept and slope to graph a linear equation.

_____

_____

_____

NAME _____ DATE _____ PERIOD _____

4-5 Scatter Plots and Lines of Fit

What You'll Learn Skim the lesson. Predict two things that you expect to learn based on the headings and the Key Concept box.

1. _____

2. _____

Active Vocabulary **New Vocabulary** Match the term with its definition by drawing a line to connect the two.

bivariate data a set of bivariate data graphed as ordered pairs on a coordinate plane

line of fit a set of data which contains two variables

scatter plot the process of using a linear equation to predict values inside the range of a set of data

linear interpolation a line which closely approximates the scatter plot for a set of data

Vocabulary Link Circle each word which would likely describe the given statistical relationship.

1. the amount of allowance and the number of CDs owned by fifteen students randomly selected from an algebra class

   negative   positive   no   weak   strong
   correlation correlation correlation correlation correlation

2. the height in inches and the number of hours spent sleeping each week for ten adults selected at random

   negative   positive   no   weak   strong
   correlation correlation correlation correlation correlation

3. the number of hours worked and the number of hours spent watching TV each week by nine teenagers selected at random

   negative   positive   no   weak   strong
   correlation correlation correlation correlation correlation

NAME _____ DATE _____ PERIOD _____

Lesson 4-5 *(continued)*

Main Idea	Details
Investigate Relationships Using Scatter Plots	Describe a real-world situation and a set of corresponding data that would show a strong positive correlation. Describe the meaning of the correlation in terms of the real-world situation. Situation: 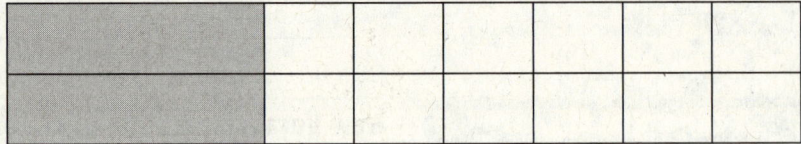 Correlation Meaning:
Use Lines of Fit	Make a scatter plot and describe the correlation. Determine a line of fit for the data. Use the line of fit to predict the number of hours exercised per week by a 15-year-old. The table shows the number of hours spent exercising per week and the age of a random sample of seven people. | Age	18	26	32	38	52	59
---	---	---	---	---	---	---
Hours	10	5	2	3	1.5	1	 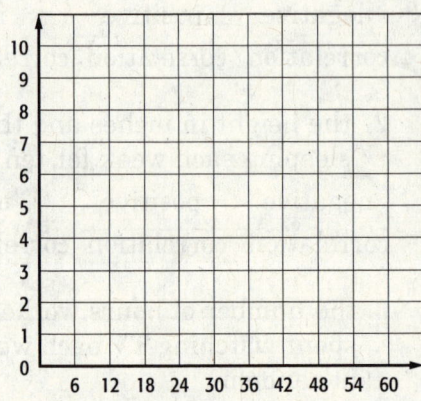 line of fit

NAME _____ DATE _____ PERIOD _____

4-6 Regression and Median-Fit Lines

What You'll Learn Scan the text under the *Now* heading. List two things you will learn about in the lesson.

1. _____

2. _____

Active Vocabulary **New Vocabulary** Write the definition next to each term.

best-fit line ▶ _____
_____
_____

linear regression ▶ _____
_____
_____

correlation coefficient ▶ _____
_____
_____

median-fit line ▶ _____
_____
_____

Vocabulary Link Consider the statement "There is a strong correlation between smoking cigarettes and developing lung cancer." Explain this statement mathematically and indicate a probable value for the correlation coefficient.

_____
_____
_____
_____
_____

Chapter 4   71   Glencoe Algebra 1

NAME _____ DATE _____ PERIOD _____

Lesson 4-6 (continued)

Main Idea	Details
Best-Fit Lines	Record the keystrokes required to perform linear regression on your calculator. Provide details as necessary. [entering the data] [performing regression] [graphing the scatter plot and regression line]
Median-Fit Lines	Use your graphing calculator to determine the median-fit line for the following set of data. Use this equation to perform both a *linear interpolation* and a *linear extrapolation*. 	Number of Ads	2	5	8	8	10	12	 	Sales ($ thousands)	2	4	7	6	9	10	 median-fit equation: 	interpolation	extrapolation	

Helping You Remember Explain how each of the following terms are related: *scatter plot*, *line of fit*, *best-fit line*, *regression line*, and *median-fit line*.

_____

_____

_____

Chapter 4   72   Glencoe Algebra 1

NAME _____ DATE _____ PERIOD _____

4-7 Inverse Linear Functions

What You'll Learn Scan the lesson. List two headings you would use to make an outline of this lesson.

1. _____

2. _____

Active Vocabulary **Review Vocabulary** Fill in the blanks with the correct term or phrase. *(Lessons 1-6 and 1-7)*

function A(n) _____ is a relationship between input and output.

mapping A(n) _____ illustrates how each element of the domain is paired with an element in the range.

ordered pair A(n) _____ is a relationship between input and output.

relation A(n) _____ is the set of numbers or coordinates used to locate any point an a coordinate plane.

New Vocabulary Write the definition next to each term.

inverse ▶ _____

_____

inverse relation ▶ _____

_____

_____

inverse function ▶ _____

_____

_____

Chapter 4   73   Glencoe Algebra 1

NAME _____ DATE _____ PERIOD _____

Lesson 4-7 (continued)

Main Idea	Details
Inverse Relations	**Graph the inverse relation of the line shown.**
	To find the inverse, exchange the coordinates of the ordered pairs.

$(-4, -7) \rightarrow$ _____

$(0, -4) \rightarrow$ _____

$(4, -1) \rightarrow$ _____

$(16, 8) \rightarrow$ _____

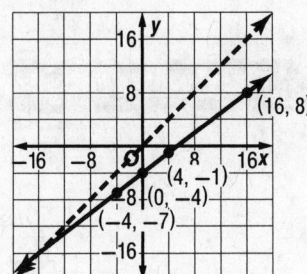

Graph these points then draw the line that passes through them

Inverse Functions

Find the inverse of $f(x) = \frac{1}{2}x - 6$

_____ Original equation

Step 1 _____ Replace $f(x)$ with y

Step 2 _____ Interchange x and y

Step 3 _____ Add 6 to each side

_____ Multiply each side by 2

Step 4 _____ Replace y with $f^{-1}(x)$

_____

Check by graphing:

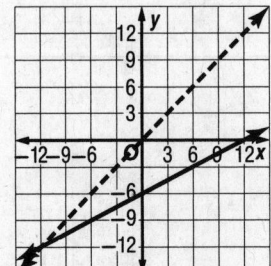

Chapter 4   74   Glencoe Algebra 1

NAME _____ DATE _____ PERIOD _____

 Chapter 4: Equations of Linear Functions

Tie It Together

Provide details for each titled graphic organizer. Supply a title and details for graphic organizers that are blank.

Equation of a Line

	Point-Slope Form	Slope-Intercept Form	Standard Form
General Equation			
Using to Graph a Line			
Using to Write the Equation of a Line			

- Positive
- Negative
- None
- Lines of Fit

[blank box]

Bivariate Data

- Scatter Plot
- Estimate Values
- Extrapolation
- Interpolation

Lines of Best Fit

- Regression Line
- Median-fit Line

Chapter 4   Glencoe Algebra 1

NAME _____ DATE _____ PERIOD _____

Equations of Linear Functions

Before the Test

Review the ideas you listed in the table at the beginning of the chapter. Cross out any incorrect information in the first column. Then complete the table by filling in the third column.

K What I know…	W What I want to find out…	L What I learned…

Visit connectED.mcgraw-hill.com to access your textbook, more examples, self-check quizzes, personal tutors, and practice tests to help you study for concepts in Chapter 4.

Are You Ready for the Chapter Test?

Use this checklist to help you study.

☐ I used my Foldable to complete the review of all or most lessons.

☐ I completed the Chapter 4 Study Guide and Review in the textbook.

☐ I took the Chapter 4 Practice Test in the textbook.

☐ I used the online resources for additional review options.

☐ I reviewed my homework assignments and made corrections to incorrect problems.

☐ I reviewed all vocabulary from the chapter and their definitions.

- Make up an invented sentence (acrostic) to remember lists or sequences. **P**lease **E**xcuse **M**y **D**ear **A**unt **S**ally is one acronym for remembering the order of operations (**p**arentheses, **e**xponents, **m**ultiply and **d**ivide, **a**dd and **s**ubtract).

Chapter 4   76   Glencoe Algebra 1

NAME _____ DATE _____ PERIOD _____

Chapter 5: Linear Inequalities

Before You Read

Before you read the chapter, respond to these statements.
1. Write an **A** if you agree with the statement.
2. Write a **D** if you disagree with the statement.

Before You Read	Linear Inequalities
	• Inequalities are solved by isolating the variable.
	• If both sides of an inequality are multiplied by a negative number, the inequality sign is reversed.
	• A graph of an inequality has an open circle when the symbol is "greater than or equal to".
	• The order of operations does not apply when solving inequalities.
	• Inequalities with absolute values are undefined.

 Construct the Foldable as directed at the beginning of this chapter.

Note Taking Tips

- **Remember to study your notes daily.**
  Reviewing small amounts at a time will help you retain the information.

- **When you take notes, it may be helpful to sit as close as possible to the front of the class.**
  There are fewer distractions and it is easier to hear.

NAME _____ DATE _____ PERIOD _____

CHAPTER 5 Linear Inequalities

Key Points

Scan the pages in the chapter and write at least one specific fact concerning each lesson. For example, in the lesson on solving inequalities by addition and subtraction, one fact might be that when solving inequalities, the goal is to isolate the variable on one side of the inequality. After completing the chapter, you can use this table to review for your chapter test.

Lesson	Fact
5-1 Solving Inequalities by Addition and Subtraction	
5-2 Solving Inequalities by Multiplication and Division	
5-3 Solving Multi-Step Inequalities	
5-4 Solving Compound Inequalities	
5-5 Inequalities Involving Absolute Value	
5-6 Graphing Inequalities in Two Variables	

Chapter 5   Glencoe Algebra 1

5-1 Solving Inequalities by Addition and Subtraction

What You'll Learn

Scan the text under the *Now* heading. List two things you will learn about in the lesson.

1. _____

2. _____

Active Vocabulary

Review Vocabulary Write a word description for each inequality symbol and write a true mathematical sentence using the symbol. *(Lesson 1-1)*

1. > _____   _____
2. < _____   _____
3. ≥ _____   _____
4. ≤ _____   _____

New Vocabulary Label the parts of the *set builder notation* below using the phrases given at the left. Show the set on the number line.

such that

the set of all numbers b

b is less than or equal to 5

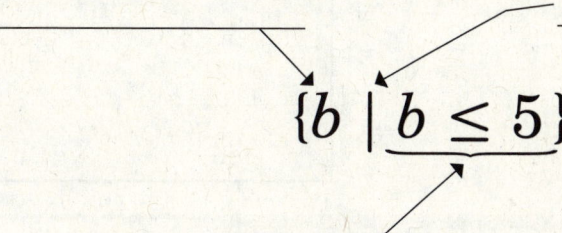

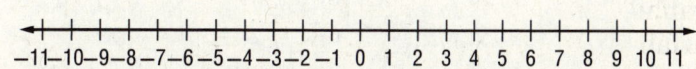

Chapter 5   79   Glencoe Algebra 1

Lesson 5-1 (continued)

Main Idea	Details
Solve Inequalities by Addition	Fill in the chart with the missing solution set representations.

Verbal Description	Set Builder Notation	Graphical Representation
all numbers greater than 3		–4 –3 –2 –1 0 1 2 3 4
		–4 –3 –2 –1 0 1 2 3 4
	$\{x \mid x < -3\}$	–4 –3 –2 –1 0 1 2 3 4

Solve Inequalities by Subtraction

Write a linear inequality to represent the following problem. Solve the inequality. Provide a complete sentence to answer the problem.

Raul needs at least $150 to purchase a digital audio player. Currently, Raul has $102. How much more money does Raul need before he can purchase a digital audio player?

Inequality:	Answer the problem:
Solution:	

Helping You Remember Teaching someone else can help you remember something. Explain how you would teach another student to solve the inequality $2x + 4 \leq 3x$.

Chapter 5   Glencoe Algebra 1

5-2 Solving Inequalities by Multiplication and Division

What You'll Learn

Scan the lesson. List two headings you would use to make an outline of this lesson.

1. _____

2. _____

Active Vocabulary

Review Vocabulary Explain how the *Multiplication Property of Equality* and the *Division Property of Equality* can both be used to solve the equation $3x = 24$. *(Lesson 2-2)*

Multiplication Property of Equality	Division Property of Equality

Vocabulary Link Solve the inequality below by following the outlined steps.

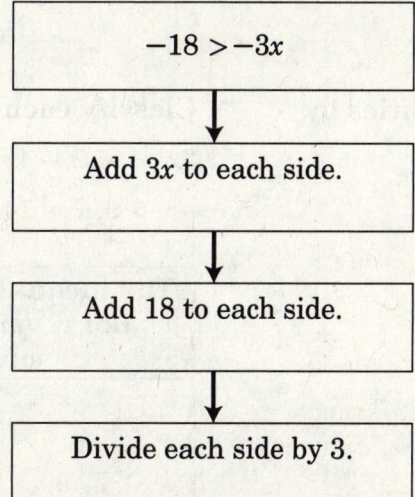

Chapter 5   81   Glencoe Algebra 1

NAME _____ DATE _____ PERIOD _____

Lesson 5-2 (continued)

Main Idea	Details
Solve Inequalities by Multiplication	Compare and contrast the process for solving the inequalities $-\frac{1}{3}x > -12$ and $\frac{1}{3}x > 12$ and for showing the solution sets on a number line. Similarities: _____ _____ _____ Differences: _____ _____ _____ _____ _____ _____ _____ _____
Solve Inequalities by Division	Classify each inequality listed in the chart below. $3x > -12$, $-4x < 15$, $-\frac{2}{3}x \leq -15$, $x - 5 > -15$, $\frac{1}{4}x \geq -8$, $-x > 9$, $x + 14 < -6$, $\frac{3}{2}x > -7$ | The inequality symbol is not reversed when solving.	The inequality symbol is reversed when solving.

Chapter 5   82   Glencoe Algebra 1

NAME _____ DATE _____ PERIOD _____

5-3 Solving Multi-Step Inequalities

What You'll Learn Skim the Examples in the lesson. Predict two things you think you will learn about solving multi-step inequalities.

1. _____

2. _____

Active Vocabulary **Review Vocabulary** Use the Distributive Property to simplify each expression. *(Lesson 1-3)*

1. $3(2x - 7)$ 2. $-4x + 2(3x + 1)$

3. $2(x + 5) + 3(2x + 1)$ 4. $-4(2x - 6) - (x + 7)$

Vocabulary Link Fill in a missing term in each equation to satisfy the given solution. Justify your answer by solving each equation.

1. $4x - 12 = 6x +$ ☐ Solution: $x = -2$

2. $2x - 10 = 2x +$ ☐ Solution: ∅

3. $3x + 11 =$ ☐ $+ 11$ Solution: {all real numbers}

Chapter 5 83 Glencoe Algebra 1

NAME _____ DATE _____ PERIOD _____

Lesson 5-3 (continued)

Main Idea	Details
Solve Multi-Step Inequalities	Solve each inequality using the indicated first step. Show the solution set using set builder notation and on a number line.

$5 - 6z \geq 13$
Subtract 5 from each side.

$5 - 6z \geq 13$
Add $6z$ to each side.

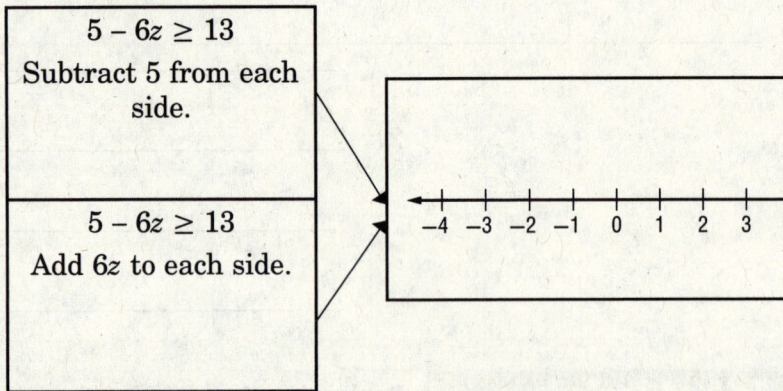

Solve Inequalities Involving the Distributive Property

Explain how to identify an inequality that has either "all real numbers" or "∅" as the solution.

All Real numbers: _____
_____
_____

∅: _____
_____
_____

Helping You Remember Make a checklist of steps for solving inequalities.

_____
_____
_____
_____

Chapter 5   84   Glencoe Algebra 1

5-4 Solving Compound Inequalities

What You'll Learn

Skim the lesson. Write two things you already know about solving compound inequalities.

1. _____

_____

2. _____

_____

Active Vocabulary

Review Vocabulary Match each verbal description to the correct inequality symbol. *(Lesson 1-1)*.

$x \geq 12$ is no more than 12

$12 < x$ is less than 12

$x \leq 12$ is at least 12

$x < 12$ is more than 12

New Vocabulary Write the correct term beside each definition.

_____ ▸ Corresponds to the word "and". Solutions are common to both inequalities in a compound inequality.

The name given to two inequalities considered together.

_____ ▸

_____ ▸ Corresponds to the word "or". Solutions are from one, the other, or both inequalities in a compound inequality.

Vocabulary Link Shade the *intersection* of sets A and B in Diagram I. Shade the *union* of sets A and B in Diagram II.

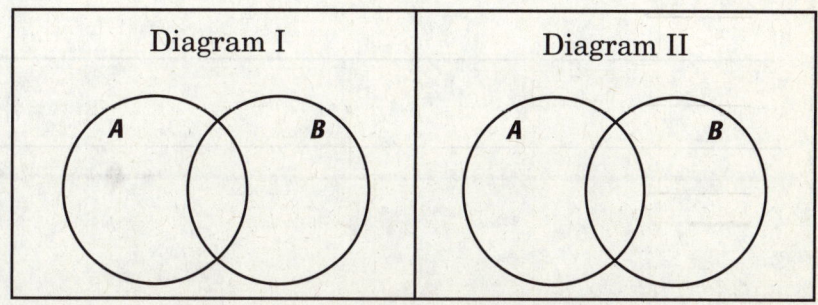

NAME _____ DATE _____ PERIOD _____

Lesson 5-4 (continued)

Main Idea	Details
Inequalities Containing *and*	Complete the diagram to solve the inequality. 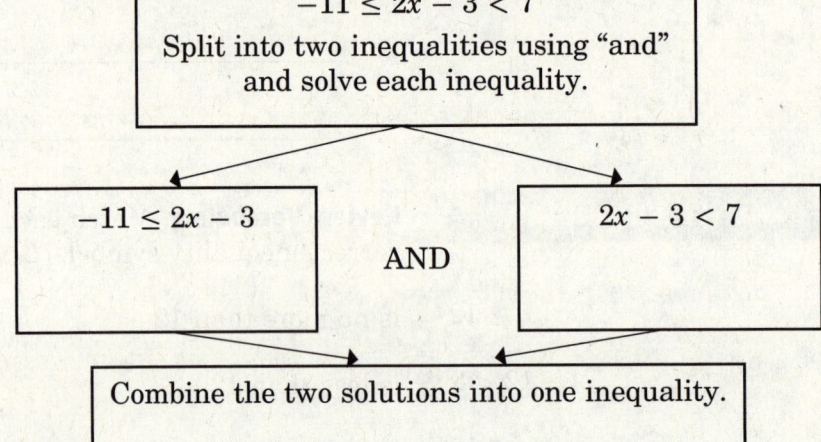
Inequalities Containing *or*	To be on the Tiny Tigers Tennis Team, a child must be at least 6 years old, but less than 9 years old. Write two *compound inequalities*: one representing the ages of children who can be on the team, and the other representing the ages of children who cannot be on the team. Children on the team: _____ Children not on the team: _____

Helping You Remember One way to remember something is to connect it to something that is familiar to you. Write two *true* compound statements about yourself, one using the word *and* and other using the word *or*.

_____

_____

Chapter 5   86   Glencoe Algebra 1

5-5 Inequalities Involving Absolute Value

What You'll Learn

Scan the text in the lesson. Write two facts you learned about inequalities involving absolute value as you scanned the text.

1. _____

2. _____

Active Vocabulary

Review Vocabulary Solve each *absolute value* equation. (Lesson 2-5)

1. $|x| = 12$

2. $|x| - 5 = -20$

3. $4|x - 6| = 16$

4. $|3x - 1| + 2 = 18$

Vocabulary Link Shade the areas on the coordinate planes which meet the conditions. Describe the shape of the shaded region.

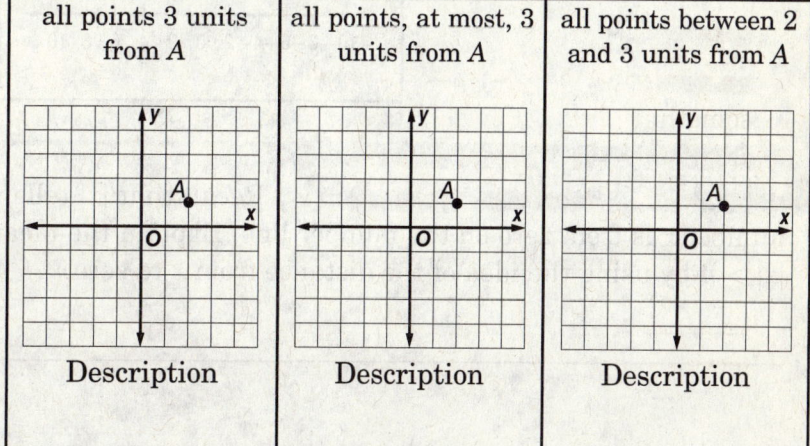

all points 3 units from A	all points, at most, 3 units from A	all points between 2 and 3 units from A
Description	Description	Description

Lesson 5-5 (continued)

Main Idea	Details
Inequalities Involving Absolute Value	Complete the chart below for solving absolute value inequalities.

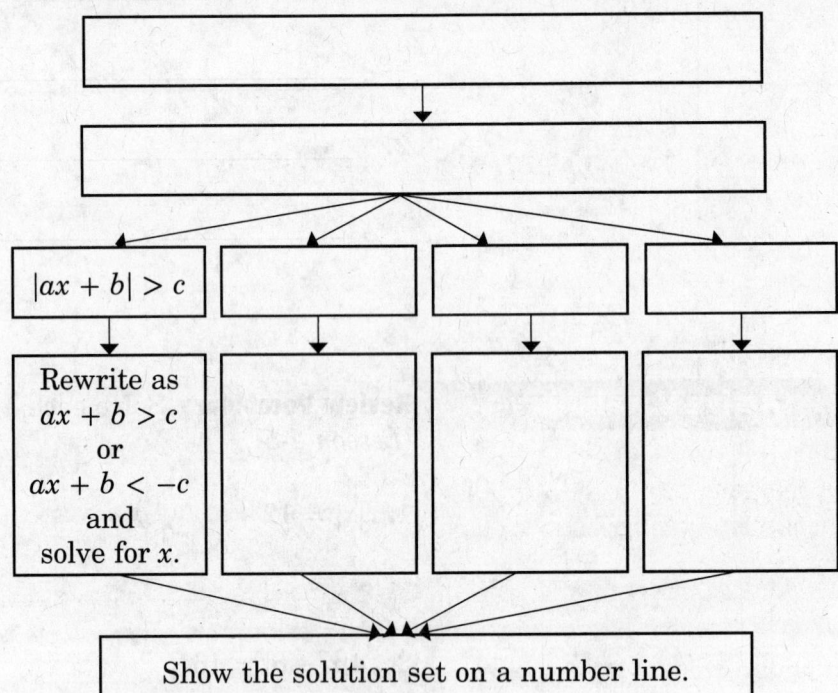

Solve each inequality.

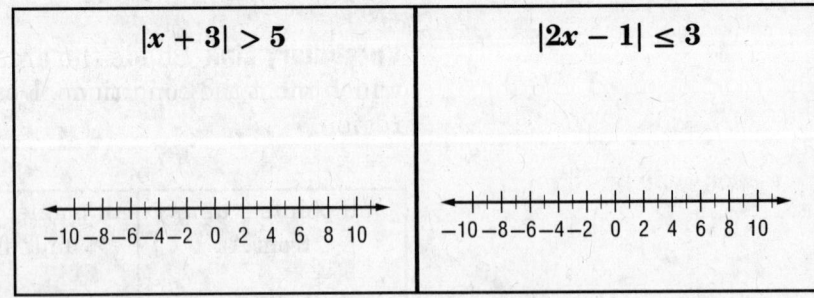

Helping You Remember

Recall that $|x|$ tells you how many units the number x is from zero on the number line. Explain the meaning of $|x| = n$, $|x| < n$ and $|x| > n$ by using the idea of the distance from x to zero.

_____

_____

5-6 Graphing Inequalities in Two Variables

What You'll Learn

Skim the lesson. Predict two things that you expect to learn based on the headings and the Key Concept box.

1. _____

2. _____

Active Vocabulary

New Vocabulary Write the definition next to each term.

boundary ▶ _____

half-plane ▶ _____

closed half-plane ▶ _____

open half-plane ▶ _____

NAME _____ DATE _____ PERIOD _____

Lesson 5-6 (continued)

Main Idea	Details
Graph Linear Inequalities	Sequence the steps for graphing a linear inequality by placing one step in each box. Add details in the box next to each step. *Shade the graph, Graph the boundary line, Determine if the boundary line is solid or shaded, Pick a point not on the line to test, Check a point not in the shaded region*

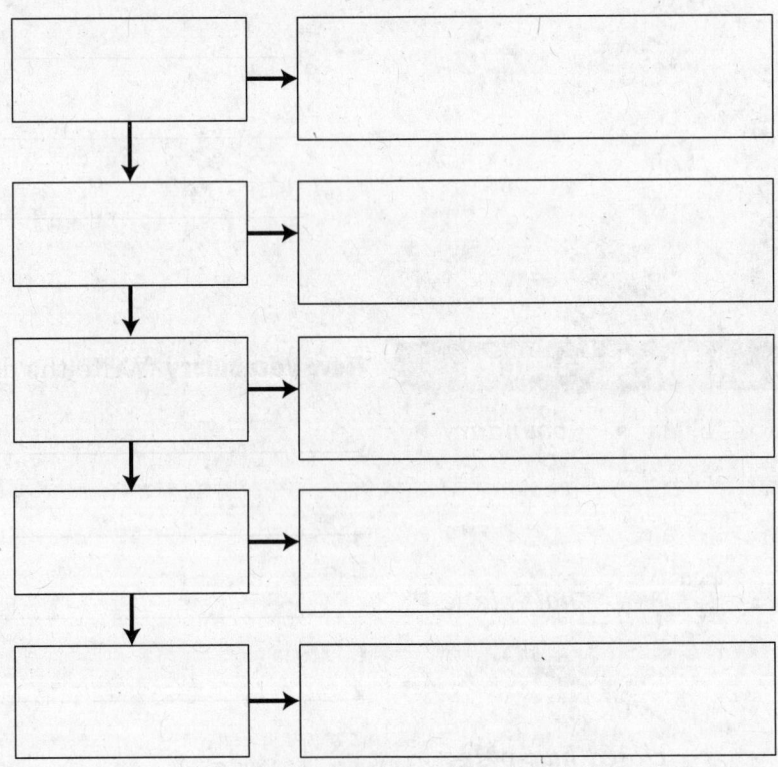

Solve Linear Inequalities	Use an inequality in two variables to solve $-2x - 3 \leq -5$.

Write the related function.

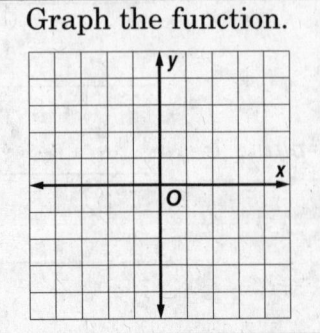

Graph the function.

Pick/Test a Point. Shade the graph.

Chapter 5   90   Glencoe Algebra 1

NAME _____ DATE _____ PERIOD _____

CHAPTER 5: Linear Inequalities

Tie It Together

Provide the indicated details in the graphic organizer.

Using Properties of Inequalities to Solve One-Step Inequalities

- Addition Example
- Subtraction Example
- Multiplication Example
- Division Example

Using Properties of Inequalities to Solve Two-Step Inequalities

Step 1:

Step 2:

Compound Inequalities

- Intersection Example
  - Absolute Value Inequalities Example
- Union Example
  - Absolute Value Inequalities Example

Chapter 5 — Glencoe Algebra 1

NAME _____ DATE _____ PERIOD _____

Linear Inequalities

Before the Test

Now that you have read and worked through the chapter, think about what you have learned and complete the table below. Compare your previous answers with these.

1. Write an **A** if you agree with the statement.
2. Write a **D** if you disagree with the statement.

Linear Inequalities	After You Read
• Inequalities are solved by isolating the variable.	
• If both sides of an inequality are multiplied by a negative number, the inequality sign is reversed.	
• A graph of an inequality has an open circle when the symbol is "greater than or equal to".	
• The order of operations does not apply when solving inequalities.	
• Inequalities with absolute values are undefined.	

Visit connectED.mcgraw-hill.com to access your textbook, more examples, self-check quizzes, personal tutors, and practice tests to help you study for concepts in Chapter 5.

Are You Ready for the Chapter Test?

Use this checklist to help you study.

☐ I used my Foldable to complete the review of all or most lessons.

☐ I completed the Chapter 5 Study Guide and Review in the textbook.

☐ I took the Chapter 5 Practice Test in the textbook.

☐ I used the online resources for additional review options.

☐ I reviewed my homework assignments and made corrections to incorrect problems.

☐ I reviewed all vocabulary from the chapter and their definitions.

• On handouts, homework, and workbooks that can be written in, underline and highlight significant information.

NAME _____ DATE _____ PERIOD _____

 # Systems of Linear Equations and Inequalities

Before You Read

Before you read the chapter, think about what you know about systems of linear equations and inequalities. List three things you already know about them in the first column. Then list three things you would like to learn about them in the second column.

K What I know...	W What I want to find out...

 Construct the Foldable as directed at the beginning of this chapter.

Note Taking Tips

- **If your instructor points out definitions or procedures from your text, write a reference page in your notes.**
  You can then write these referenced items in their proper place in your notes after class.

- **When you take notes, listen or read for main ideas.**
  Then record concepts, define terms, write statements in if-then form, and write paragraph proofs.

Chapter 6   93   *Glencoe Algebra 1*

NAME _____ DATE _____ PERIOD _____

CHAPTER 6: Systems of Linear Equations and Inequalities

Key Points

Scan the pages in the chapter and write at least one specific fact concerning each lesson. For example, in the lesson on graphing systems of equations, one fact might be that if a consistent system has an infinite number of solutions, it is dependent. After completing the chapter, you can use this table to review for your chapter test.

Lesson	Fact
6-1 Graphing Systems of Equations	
6-2 Substitution	
6-3 Elimination Using Addition and Subtraction	
6-4 Elimination Using Multiplication	
6-5 Applying Systems of Linear Equations	
6-6 Systems of Inequalities	

Chapter 6   94   Glencoe Algebra 1

6-1 Graphing Systems of Equations

What You'll Learn

Scan the text under the *Now* heading. List two things you will learn about in the lesson.

1. _____

_____

2. _____

_____

Active Vocabulary

Review Vocabulary Make a table of values which satisfy the equation $x + y = 13$. *(Lesson 3-1)*

x									
y									

Is it possible to make a table that shows all ordered pairs that satisfy this equation? Justify your answer.

_____

_____

_____

How can you show all of the ordered pairs for the equation?

_____

_____

New Vocabulary Match the term with its definition by drawing a line.

consistent — a set of two or more equations that contain the same variables

inconsistent — a system of equations that has at least one solution

system of equations — a system of equations that has an infinite number of solutions

independent — a system of equations that has exactly one solution

dependent — a system of equations that has no solutions

NAME _____ DATE _____ PERIOD _____

Lesson 6-1 (continued)

Main Idea	Details
Possible Number of Solutions	Add a line to each graph so that the given condition is satisfied. 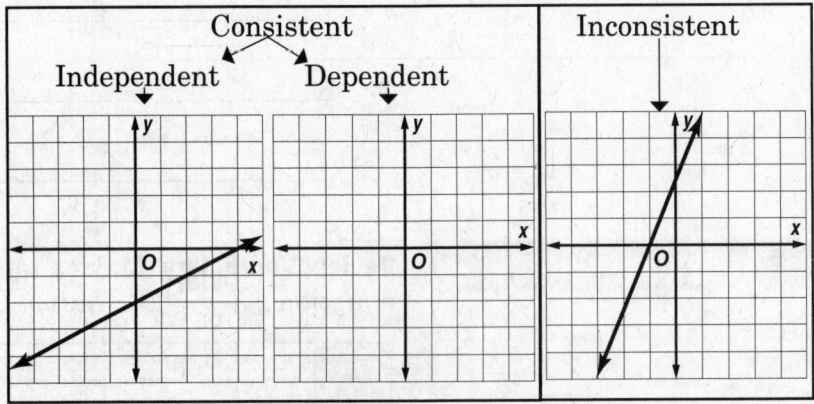
Solve by Graphing	Solve the system of equations by graphing. 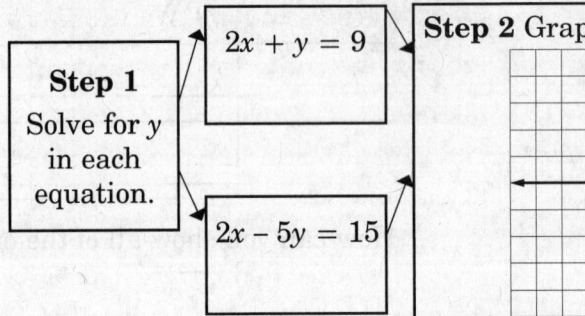 Step 2 Graph each equation. Step 3 Find the solution. The lines intersect at point _____.

Helping You Remember Describe how you can solve a system of equations by graphing.

_____

_____

Chapter 6   96   Glencoe Algebra 1

NAME _____ DATE _____ PERIOD _____

6-2 Substitution

What You'll Learn

Scan the text in the lesson. Write two facts you learned about solving systems by substitution as you scanned the text.

1. _____

_____

2. _____

_____

Active Vocabulary

Review Vocabulary Solve the equation after substituting the given value for each variable. *(Lesson 2-3)*

1. $3x + 7y = 8$, given $x = -2$
2. $-2y + 2x = 12$, given $y = 0$
3. $y - \frac{2}{3}x = 9$, given $x = -6$
4. $0.5y + 6x = -5$, given $y = 4$

New Vocabulary Write the definition next to the term.

substitution ▶ _____

_____

_____

_____

Vocabulary Link Describe when it would be more convenient to use substitution than graphing for solving a system of equations.

_____

_____

_____

Chapter 6   97   Glencoe Algebra 1

NAME _____ DATE _____ PERIOD _____

Lesson 6-2 (continued)

Main Idea	Details
Solve by Substitution	Solve the system of equations twice using the *substitution method*. In the first column, solve for x initially. In the second, solve for y initially. 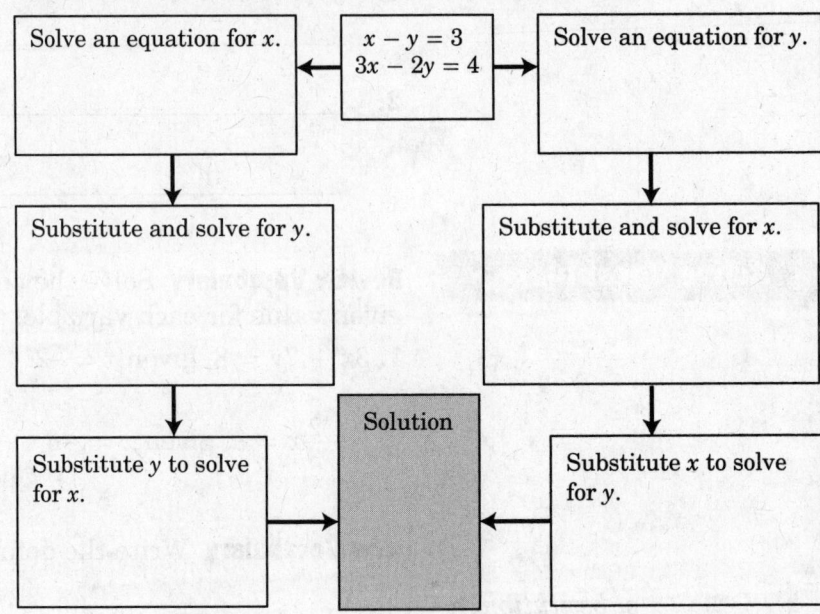
Solve Real-World Problems	Write a system of equations to represent the following problem. Identify the variables. Solve the system. A total of 150 tickets were sold for the annual concert. Student tickets were \$4 and non-student tickets were \$8. If the total revenue was \$840, how many tickets of each type were sold?

Let $s =$	system	Solve and answer.
Let $n =$		

Helping You Remember

What is usually the first step in solving a system of equations by substitution?

_____

_____

Chapter 6   98   Glencoe Algebra 1

6-3 Elimination Using Addition and Subtraction

What You'll Learn

Scan the lesson. List two headings you would use to make an outline of this lesson.

1. _____

_____

2. _____

_____

Active Vocabulary

Review Vocabulary Match each linear equation with the appropriate form. *(Lessons 4-2 and 4-3)*

slope-intercept form $\quad y = -\frac{3}{4}x + 3$

point-slope form $\quad 3x + 4y = 12$

standard form $\quad y + 3 = (x - 8)$

Do these equations represent the same line? Justify your answer.

_____

_____

_____

New Vocabulary Fill in the blanks with the correct term or phrase.

elimination ▶ It is a method to _____ a system in which the equations are written so that like _____ with the same or opposite coefficients are _____. The equations are _____ or subtracted to eliminate one _____. The value for one variable is found and is _____ into one of the equations to solve for the other variable.

Chapter 6   99   Glencoe Algebra 1

Lesson 6-3 (continued)

Main Idea	Details

Elimination Using Addition

Solve each system of equations using the addition method. Fill in both the verbal and mathematical missing steps.

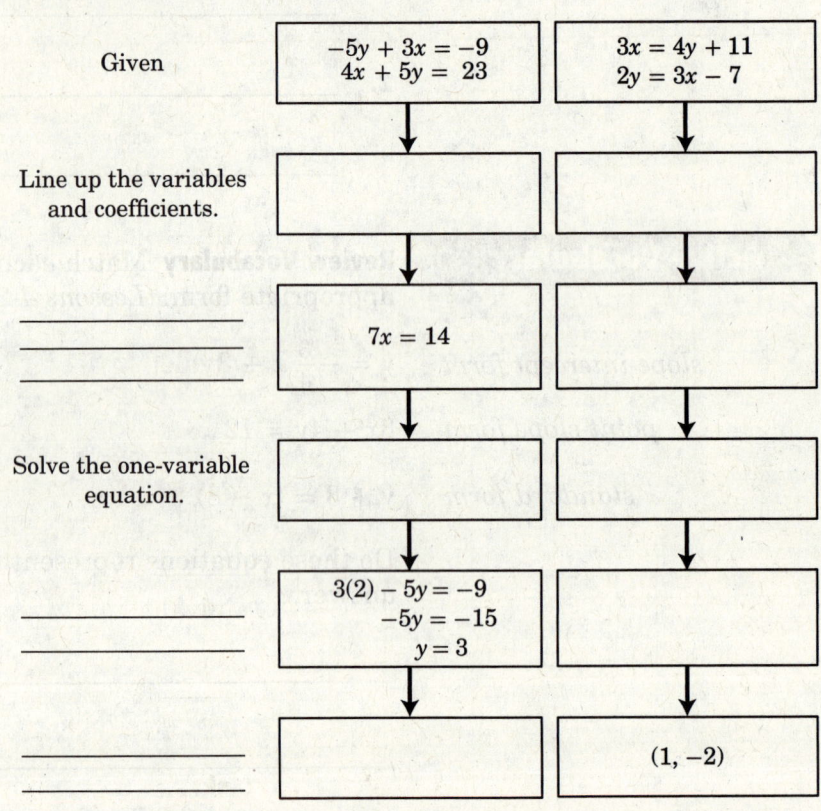

Elimination Using Subtraction

Create a system of equations which has a solution of (2, 4) and can be solved using the subtraction method.

Helping You Remember

Tell how you can decide whether to use addition or subtraction to eliminate a variable in a system of equations.

NAME _____ DATE _____ PERIOD _____

6-4 Elimination Using Multiplication

What You'll Learn Skim the lesson. Predict two things that you expect to learn based on the headings and the Key Concept box.

1. _____

_____

2. _____

_____

Active Vocabulary **Review Vocabulary** Write the property of equality which is represented by each example. *(Lesson 1-3)*

$4x = 9$ is equivalent to $4x - 18 = -9$.

_____

$3x + 2y = 12$ is equivalent to $6x + 4y = 24$.

_____

$3x = 12$ is equivalent to $3x + 8 = 20$.

_____

Vocabulary Link Add the two linear equations to create a third. Graph all three equations on the same plane. What happens?

_____

_____

$2x - 3y = -8$
$-x + 2y = 6$

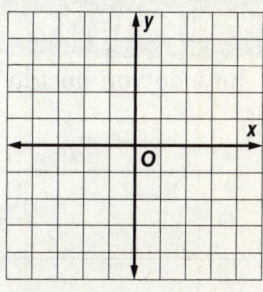

Chapter 6　　　101　　　Glencoe Algebra 1

Lesson 6-4 (continued)

Main Idea	Details

Elimination Using Multiplication

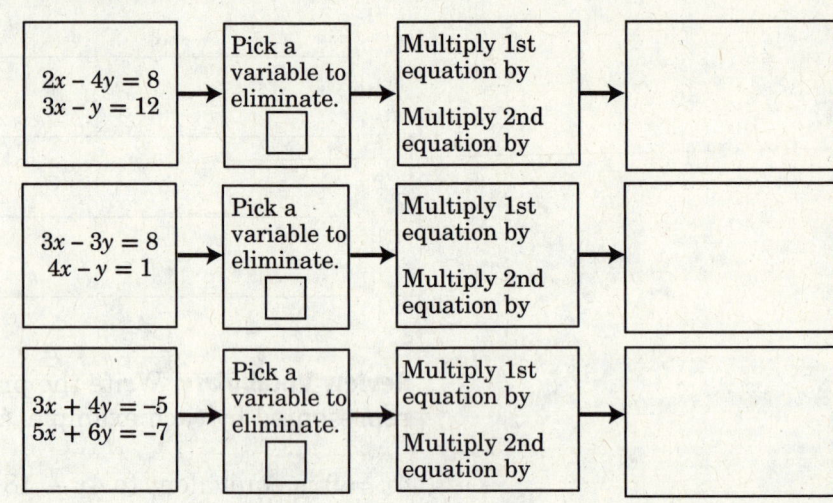

Solve Real-World Problems

Write a system of equations to represent the following problem. Identify the variables. Solve the system using elimination.

On Monday, Arnold paid $3.40 for three donuts and two coffees. On Tuesday, he paid $3.60 for two donuts and three coffees. On Wednesday, he bought one donut and one coffee. What was his bill for one donut and one coffee?

Let $d =$	system	Solve and answer.
Let $c =$	$3d + 2c = 3.40$ $2d + 3c = 3.60$	

Helping You Remember If you are going to solve a system by elimination, how do you decide whether you will need to multiply one or both equations by a number?

_____

_____

_____

6-5 Applying Systems of Linear Equations

What You'll Learn Skim the Examples in the lesson. Predict two things you think you will learn about applying systems of equations.

1. _____
   _____
   _____
   _____

2. _____
   _____
   _____
   _____

Active Vocabulary **Review Vocabulary** Solve the system of equations using each of the four methods. *(Lessons 6-1 through 6-4)*

$$x - 2y = 4; \quad x - y = 3$$

Graphing	Substitution
Solution:	Solution:
Elimination Using Subtraction	Elimination Using Multiplication
_____	_____
Solution:	Solution:

Lesson 6-5 (continued)

Main Idea	Details
Determine the Best Method	Summarize when to use each of the following methods in your own words. 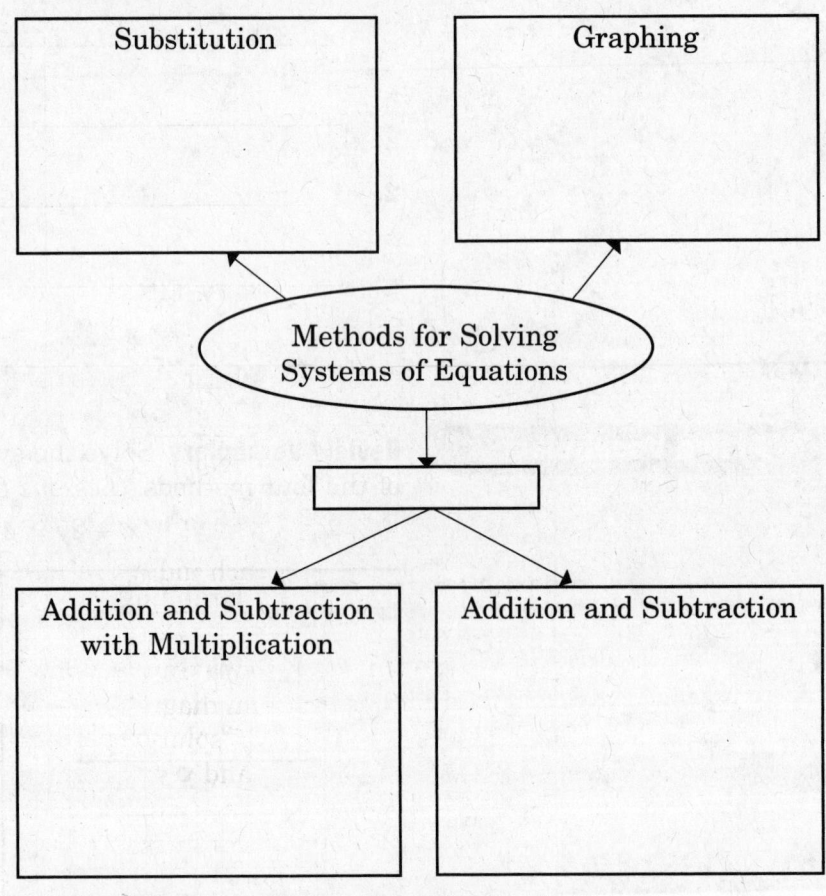
Apply Systems of Linear Equations	Write a word problem that could be represented by the following system of equations. Identify each variable. $4b + 3m = 1.45$; $2b + 5m = 1.25$ $b = $ $m = $ Word Problem

NAME _____ DATE _____ PERIOD _____

6-6 Systems of Inequalities

What You'll Learn Scan the text under the *Now* heading. List two things you will learn about in the lesson.

1. _____

    _____

2. _____

    _____

Active Vocabulary

New Vocabulary Fill in the blank with the correct term or phrase.

system of inequalities ▶

It is a set of _____ or more inequalities with the same _____. The solution of the system is the set of _____ that satisfy all of the inequalities in the system. These ordered pairs are the _____ of the graphs of each individual inequality.

Vocabulary Link Intersecting regions can be represented using a Venn diagram. Place the terms "solutions of $y > 2x - 4$", "solutions of $y \leq -0.5x + 3$", and "solutions of $y > 2x - 4$ and $y \leq -0.5x + 3$" in the Venn diagram below.

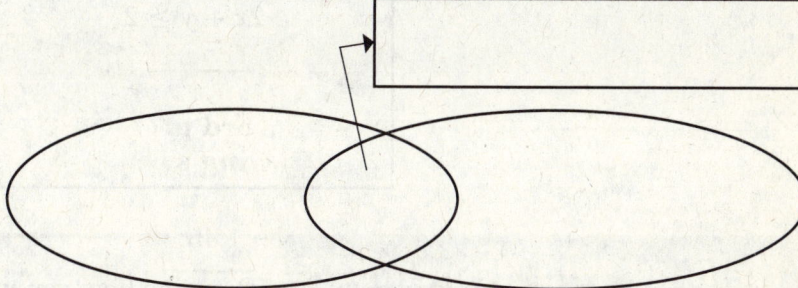

How could you use the Venn diagram to represent "solutions of $y > 2x - 4$ or $y \leq -0.5x + 3$"?

_____

_____

NAME _____ DATE _____ PERIOD _____

Lesson 6-6 *(continued)*

Main Idea	Details
Systems of Inequalities	**Sequence the steps for solving a system of inequalities. Solve the two systems of inequalities.** *Graph the first inequality and shade appropriately. Write both inequalities in slope-intercept form. Determine the intersection(s) of the shaded regions. Graph the second inequality and shade appropriately.*

☐ → ☐ → ☐ → ☐

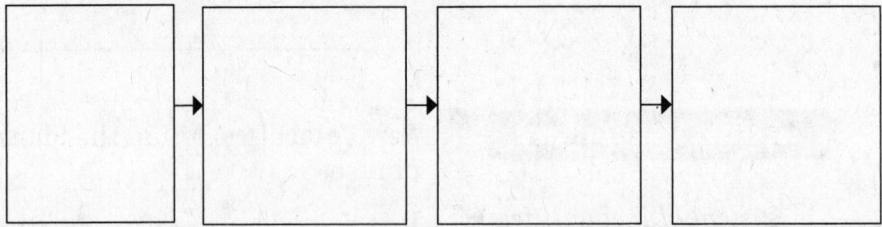

Example 1	
$y > x - 1$ $y < x + 3$	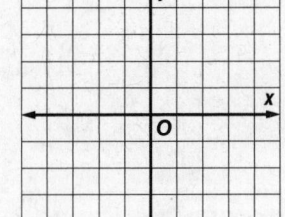
Example 2	
$4x - 3y \leq 3$ $2x + y \geq 2$	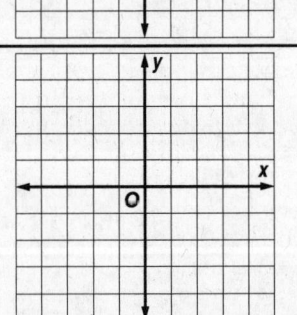

Helping You Remember Describe how you would explain the process of using a graph to solve a system of inequalities to a friend.

_____
_____

Chapter 6   106   Glencoe Algebra 1

NAME _____ DATE _____ PERIOD _____

CHAPTER 6: Systems of Linear Equations and Inequalities

Tie It Together

Fill in each graphic organizer. Add details if space permits.

Solving Systems of Equations

Method	How to Use	When to Use
Graphing		
Substitution		
Elimination with Addition/ Subtraction		
Elimination with Multiplication		

Possible Solution Sets

Algebraically Solved	Algebraically Solved	Algebraically Solved
Graphically Solved	Graphically Solved	Graphically Solved

Chapter 6 107 Glencoe Algebra 1

NAME _____ DATE _____ PERIOD _____

CHAPTER 6
Systems of Linear Equations and Inequalities

Before the Test

Review the ideas you listed in the table at the beginning of the chapter. Cross out any incorrect information in the first column. Then complete the table by filling in the third column.

K What I know…	W What I want to find out…	L What I learned…

Visit connectED.mcgraw-hill.com to access your textbook, more examples, self-check quizzes, personal tutors, and practice tests to help you study for concepts in Chapter 6.

Are You Ready for the Chapter Test?

Use this checklist to help you study.

☐ I used my Foldable to complete the review of all or most lessons.

☐ I completed the Chapter 6 Study Guide and Review in the textbook.

☐ I took the Chapter 6 Practice Test in the textbook.

☐ I used the online resources for additional review options.

☐ I reviewed my homework assignments and made corrections to incorrect problems.

☐ I reviewed all vocabulary from the chapter and their definitions.

 Study Tip

- Use the SQ3R method of reading: **S**urvey, **Q**uestion, **R**ead, **R**ecite, and **R**eview. Survey the text by previewing the headings, boldface words, and examples. Ask questions about what you survey, read with purpose, recite out loud the main points and concepts without looking at the text, and review your text notes or use the chapter review at the end of the chapter.

Chapter 6   108   Glencoe Algebra 1

NAME _____ DATE _____ PERIOD _____

Chapter 7: Exponents and Exponential Functions

Before You Read

Before you read the chapter, respond to these statements.
1. Write an **A** if you agree with the statement.
2. Write a **D** if you disagree with the statement.

Before You Read	Exponents and Exponential Functions
	• To multiply exponents with the same base, find the product of the base and the exponents.
	• A simplified expression is without fractions, duplicate bases, and powers of powers.
	• A base with a negative exponent is written with a positive exponent when it is a denominator.
	• Compound interest problems are an example of an exponential growth function.
	• In an exponential function, the base is a variable and the exponent is a constant.

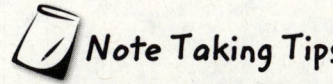 Construct the Foldable as directed at the beginning of this chapter.

Note Taking Tips

- **When taking notes, writing a paragraph that describes the concepts, the computational skills, and the graphics will help you to understand the math in the lesson.**

- **Before each lesson, skim through the lesson and write any questions that come to mind in your notes.**

As you work through the lesson, record the answer to your question.

NAME _____ DATE _____ PERIOD _____

Exponents and Exponential Functions

Key Points

Scan the pages in the chapter and write at least one specific fact concerning each lesson. For example, in the lesson on division properties of exponents, one fact might be that the order of magnitude of a quantity is the number rounded to the nearest power of 10. After completing the chapter, you can use this table to review for your chapter test.

Lesson	Fact
7-1 Multiplication Properties of Exponents	
7-2 Division Properties of Exponents	
7-3 Rational Exponents	
7-4 Scientific Notation	
7-5 Exponential Functions	
7-6 Growth and Decay	
7-7 Geometric Sequences as Exponential Functions	
7-8 Recursive Formulas	

7-1 Multiplication Properties of Exponents

What You'll Learn

Skim the Examples in the lesson. Predict two things you think you will learn about multiplication properties of exponents.

1. _____

2. _____

_____

_____

Active Vocabulary

New Vocabulary Fill in each blank with the correct term or phrase.

constant ▶ A constant is a monomial that is a _____.

monomial ▶ A monomial is a number, a _____, or the product of a number and one or more variables with nonnegative integer _____.

Vocabulary Link The word *constant* has a place in a number of real-world applications. Think of a real-world example where you would describe something as constant. Then look up the word and explain how its everyday meaning relates to its mathematical meaning.

_____

_____

_____

_____

Lesson 7-1 (continued)

Main Idea	Details

Multiplying Monomials

Complete the table by circling the property of powers that can be used to simplify each expression. Then simplify the expression.

Expression	Property	Simplified Expression
$(w^3)^5$	Product of Powers Power of a Power Power of a Product	
$c^2 \cdot c^4$	Product of Powers Power of a Power Power of a Product	
$(2mn)^3$	Product of Powers Power of a Power Power of a Product	

Simplify Expressions

Simplify each expression.

1. $(2mn^2)^2 (3m^2n^4)^3$

2. $(4c^2d^3)^2 [(-3c^2d^4)^3]^2$

Helping You Remember Write an example of each of the three properties of powers discussed in this lesson. Then, using the examples, explain how the property is used to simplify them.

Chapter 7   112   Glencoe Algebra 1

7-2 Division Properties of Exponents

What You'll Learn

Skim the lesson. Write two things you already know about dividing monomials.

1. _____
   _____
   _____

2. _____
   _____
   _____

Active Vocabulary

New Vocabulary Write the correct term next to each definition or expression.

_____ ▸ $\left(\dfrac{c}{5}\right)^0 = 1$

_____ ▸ for a given quantity, the number rounded to the nearest power of 10

_____ ▸ $a^{-2} = \dfrac{1}{a^2}$

Vocabulary Link Look up the definition of *magnitude*. Tell how the meaning compares to the order of magnitude of a quantity.

_____
_____
_____

Lesson 7-2 (continued)

Main Idea	Details
Quotients of Monomials	Complete the table by circling the property of powers that can be used to simplify each expression. Then simplify the expression.

Expression	Property	Simplified Expression
$\dfrac{a^2 b^4}{ab^2}$	Quotient of Powers / Power of a Quotient	
$\left(\dfrac{4z^3}{5}\right)^2$	Quotient of Powers / Power of a Quotient	

Simplify Expressions — Simplify each expression. Assume that no denominator is equal to zero.

1. $\left(\dfrac{7c^2 d^5}{21c^3 d^2}\right)^0$

2. $\dfrac{(m^{-1} n^3)^{-4}}{m^3 n^3}$

_____   _____

Helping You Remember Describe how you would help a friend who needs to simplify the expression $\dfrac{4x^2}{2x^5}$.

_____

_____

_____

7-3 Rational Exponents

What You'll Learn Skim the lesson. Write two things you already know about rational exponents.

1. _____

2. _____

Active Vocabulary

Review Vocabulary Match the term with its example by drawing a line to connect the two. *(Lessons 7-1 and 7-2)*

exponent x^{-3}

zero exponent x^0

negative exponent 2^x

New Vocabulary Write the definition next to each term.

rational exponent ▶ _____

cube root ▶ _____

nth root ▶ _____

exponential equation ▶ _____

Chapter 7   115   Glencoe Algebra 1

Lesson 7-3 (continued)

Main Idea	Details
Rational Exponents	Write each expression in radical form, or write each radical in exponential form. Find the first five terms of each sequence.

$\sqrt{225}$	$\sqrt{289x^6}$
$(81xy)^{\frac{1}{2}}$	$2197^{\frac{2}{3}}$

Solve Exponential Equations

Solve each equation.
$5^{5x} = 3125$

_____ Original equation

_____ Rewrite 3125 as _____.

_____ Property of Equality

_____ Divide each side by _____.

$3^{3x+3} = 6561$

_____ Original equation

_____ Rewrite 6561 as _____.

_____ Property of Equality

_____ Subtract _____ from each side.

_____ Divide each side by _____.

Helping You Remember A good way to remember a new concept is to explain it in your own words. How would you explain how to solve an exponential equation?

_____

_____

NAME _____ DATE _____ PERIOD _____

7-4 Scientific Notation

What You'll Learn Scan the text under the *Now* heading. List two things you will learn about in the lesson.

1. _____

2. _____

Active Vocabulary **Review Vocabulary** Write the definition next to each term. *(Lesson 7-1)*

constant ▶ _____

monomial ▶ _____

New Vocabulary Write the definition of the term.

scientific notation ▶ _____

NAME _____ DATE _____ PERIOD _____

Lesson 7-4 (continued)

Main Idea	Details
Scientific Notation	Follow the steps below to write 5.18×10^7 in standard form.

Step 1: Identify the exponent. $n = $ _____

Step 2: Move the decimal point n places to the right. $5.18 \times 10^7 \rightarrow$ _____

Step 3: Rewrite using commas. $5.18 \times 10^7 =$ _____

Products and Quotients in Scientific Notation

Evaluate each expression. Express the results in both scientific notation and standard form.

1. $(1.3 \times 10^{-6})(5.2 \times 10^8)$

2. $\dfrac{2.04 \times 10^9}{1.2 \times 10^{13}}$

Helping You Remember   A good way to remember a mathematical concept is to explain it to someone else. How would you tell a friend to write the decimal 0.00000012 using scientific notation?

_____

_____

_____

Chapter 7   118   Glencoe Algebra 1

NAME _____ DATE _____ PERIOD _____

7-5 Exponential Functions

What You'll Learn

Scan the text under the *Now* heading. List two things you will learn about in the lesson.

1. _____

    _____

    _____

2. _____

    _____

    _____

Active Vocabulary

New Vocabulary Write the definition next to the term.

exponential function ▶ _____

_____

_____

exponential growth function ▶ _____

_____

_____

exponential decay function ▶ _____

_____

_____

NAME _____ DATE _____ PERIOD _____

Lesson 7-5 (continued)

Main Idea	Details
Graph Exponential Functions	Complete the following table of function values and use it to help you graph the exponential function $y = 2^x$.

x	2^x	y

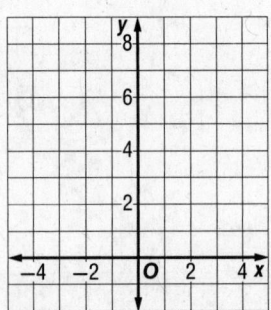

Identify Exponential Behavior

Determine whether the set of data shown below displays exponential behavior. Write *yes* or *no*. Explain why or why not.

x	0	2	4	6	8	10
y	128	64	32	16	8	4

_____

_____

_____

_____

_____

NAME _____ DATE _____ PERIOD _____

7-6 Growth and Decay

What You'll Learn Skim the Examples in the lesson. Predict two things you think you will learn about growth and decay.

1. _____

_____

_____

2. _____

_____

_____

Active Vocabulary **New Vocabulary** Fill in each blank with the correct term or phrase.

compound interest ▶ Compound interest is interest _____ or _____ on both the initial investment and previously _____ interest.

exponential decay ▶ In exponential decay, the original _____ decreases by the same _____ over a period of time.

exponential growth ▶ In exponential growth, the original amount _____ by the same percent over a period of time.

Vocabulary Link Think of some real-world examples that involve exponential growth and decay.

_____

_____

_____

Lesson 7-6 *(continued)*

Main Idea	Details
Exponential Growth	Label each of the parts of the general equation for exponential growth shown below. Use the terms *initial amount, final amount, time,* and *growth rate*.

$$y = a(1 + r)^t$$

Exponential Decay

Suppose a particular species of bird on an island is decreasing at an annual rate of 5.4%. The species originally had a population of 12,600.

a. Write an equation to represent the decrease in population.

_____

b. Estimate the number of birds on the island after 4 years.

_____

Helping You Remember A good way to help you remember a new concept is to explain it in your own words. The general equations for exponential growth and exponential decay are very similar. Explain how you can determine if an equation represents exponential growth or exponential decay.

_____

_____

_____

Chapter 7   Glencoe Algebra 1

NAME _____ DATE _____ PERIOD _____

7-7 Geometric Sequences as Exponential Functions

What You'll Learn

Skim the lesson. Write two things you already know about geometric sequences as exponential functions.

1. _____
   _____
   _____
   _____
   _____

2. _____
   _____
   _____
   _____
   _____

Active Vocabulary

New Vocabulary Write the definition next to each term.

common ratio ▶ _____
_____
_____

geometric sequence ▶ _____
_____
_____

NAME _____ DATE _____ PERIOD _____

Lesson 7-7 (continued)

Main Idea	Details
Recognize Geometric Sequences	Determine whether each sequence is *arithmetic*, *geometric*, or *neither*. Explain. 1. 12, 9, 6, 3, 0, … _____ _____ 2. 3, −6, 12, −24, 48, … _____ _____
Geometric Sequences and Functions	What is the 12th term of the geometric sequence 6, 12, 24, 48,…? **Step 1:** Compare consecutive terms to find the common ratio. $r =$ _____ **Step 2:** Write an equation to model the sequence. $a_n =$ _____ **Step 3:** Evaluate the formula for $n = 12$. $a_{12} =$ _____

Chapter 7   124   Glencoe Algebra 1

NAME _____ DATE _____ PERIOD _____

7-8 Recursive Formulas

What You'll Learn

Scan the text in the lesson. Write two facts you learned about recursive formulas as you scanned the text.

1. _____
   _____
   _____

2. _____
   _____
   _____

Active Vocabulary

New Vocabulary Write the definition next to each term.

recursive formula ▶ _____
_____
_____

explicit formula ▶ _____
_____
_____

Vocabulary Link Think of some real-world examples that involve recursive formulas.

_____
_____

Chapter 7   Glencoe Algebra 1

NAME _____ DATE _____ PERIOD _____

Lesson 7-8 (continued)

Main Idea	Details
Use Recursive Formulas	**Find the first five terms of each sequence.** 1. $a_1 = 7$ and $a_n = 3a_{n-1} - 12$, if $n \geq 2$ _____ _____ 2. $a_1 = 3$ and $a_n = 2a_{n-1} + 13$, if $n \geq 2$ _____ _____
Write Recursive Formulas	**What is the recursive formula for 3, −6, 21, −60, 183, …?** **Step 1:** First divide each term by the term that precedes it. $r = $ _____ **Step 2:** Use the formula for a geometric sequence. $a_n = $ _____ $a_1 = $ _____ **Step 3:** Identify the first term and restriction on n. n _____

Helping You Remember A good way to remember a new concept is to explain it in your own words. Explain the difference between the recursive formula and explicit formula. When is one form preferred over the other?

_____

_____

NAME _____ DATE _____ PERIOD _____

 **CHAPTER 7 Exponents and Exponential Functions**

Tie It Together

Fill in the graphic organizer with details from the chapter.

Laws of Exponents/Powers

Law	Notation	Verbal Description	Example
Product of Powers			
Power of Powers			
Power of a Product			
Quotient of Powers			
Power of a Quotient			
Zero Exponent Property			
Negative Exponent Property			

Chapter 7   127   Glencoe Algebra 1

NAME _____ DATE _____ PERIOD _____

CHAPTER 7: Exponents and Exponential Functions

Before the Test

Now that you have read and worked through the chapter, think about what you have learned and complete the table below. Compare your previous answers with these.

1. Write an **A** if you agree with the statement.
2. Write a **D** if you disagree with the statement.

Exponents and Exponential Functions	After You Read
• To multiply exponents with the same base, find the product of the base and the exponents.	
• A simplified expression is without fractions, duplicate bases, and powers of powers.	
• A base with a negative exponent is written with a positive exponent when it is a denominator.	
• Compound interest problems are an example of an exponential growth function.	
• In an exponential function, the base is a variable and the exponent is a constant.	

Visit **connectED.mcgraw-hill.com** to access your textbook, more examples, self-check quizzes, personal tutors, and practice tests to help you study for concepts in Chapter 7.

Are You Ready for the Chapter Test?

Use this checklist to help you study.

☐ I used my Foldable to complete the review of all or most lessons.

☐ I completed the Chapter 7 Study Guide and Review in the textbook.

☐ I took the Chapter 7 Practice Test in the textbook.

☐ I used the online resources for additional review options.

☐ I reviewed my homework assignments and made corrections to incorrect problems.

☐ I reviewed all vocabulary from the chapter and their definitions.

Study Tip

• Use flash cards to study for tests by writing the concept on one side of the card and its definition on the other.

NAME _____ DATE _____ PERIOD _____

Quadratic Expressions and Equations

Before You Read

Before you read the chapter, think about what you know about factoring and quadratic equations. List three things you already know about them in the first column. Then list three things you would like to learn about them in the second column.

K What I know...	W What I want to find out...

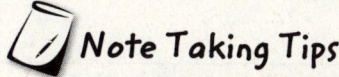 Construct the Foldable as directed at the beginning of this chapter.

Note Taking Tips

- When you take notes, always write clear and concise notes so they can be easily read when studying for a quiz or exam.

- A visual study guide like the Foldable shown above helps you organize what you know and remember what you have learned.

    You can use them to review main ideas or keywords.

Chapter 8   Glencoe Algebra 1

NAME _____ DATE _____ PERIOD _____

CHAPTER 8: Quadratic Expressions and Equations

Key Points

Scan the pages in the chapter and write at least one specific fact concerning each lesson. For example, in the lesson on multiplying polynomials, one fact might be that the FOIL method can be used as a short cut for the distributive property. After completing the chapter, you can use this table to review for your chapter test.

Lesson	Fact
8-1 Adding and Subtracting Polynomials	
8-2 Multiplying a Polynomial by a Monomial	
8-3 Multiplying Polynomials	
8-4 Special Products	
8-5 Using the Distributive Property	
8-6 Solving $x^2 + bx + c = 0$	
8-7 Solving $ax^2 + bx + c = 0$	
8-8 Differences of Squares	
8-9 Perfect Squares	

NAME _____ DATE _____ PERIOD _____

8-1 Adding and Subtracting Polynomials

What You'll Learn Skim the lesson. Predict two things that you expect to learn based on the headings and figures in the lesson.

1. _____

_____

_____

2. _____

_____

_____

Active Vocabulary **New Vocabulary** Write the correct term next to each definition.

_____ ▶ a monomial or the sum or difference of monomials, each called a *term*

_____ ▶ the coefficient of the first term of a polynomial when written in standard form

_____ ▶ the sum of the exponents of all the variables of a monomial

_____ ▶ the sum or difference of three monomials

_____ ▶ the form of a polynomial that is written with the terms in order from greatest degree to least degree

_____ ▶ the sum or difference of two monomials

_____ ▶ the greatest degree of any term in a polynomial

Chapter 8   131   Glencoe Algebra 1

Lesson 8-1 (continued)

Main Idea	Details
Polynomials in Standard Form	Complete the table below for each monomial, binomial, or trinomial.

Expression	degree	Monomial, Binomial, or Trinomial?
$32x^2y$		
$4x + 2y - 6$		
$9x^2 - 81y^2$		
-7		
$8y + 3$		

Add and Subtract Polynomials

Find each sum or difference.

1. $(3x^2 + 8) + (4x^2 - 6x)$ _____

2. $(-x^3 + 5x) - (2x^3 + 10x)$ _____

3. $(4x^2 - x + 2) + (x^2 - 3x - 8)$ _____

4. $(3x^4 + 2x^2 + 1) - (x^3 - 5x - 4)$ _____

Helping You Remember Use a dictionary to find the meaning of the terms *ascending* and *descending*. Write their meanings and then describe a situation in your everyday life that relates to them.

Chapter 8   132   Glencoe Algebra 1

NAME _____ DATE _____ PERIOD _____

8-2 Multiplying a Polynomial by a Monomial

What You'll Learn Scan the lesson. List two headings you would use to make an outline of the lesson.

1. _____

2. _____

Active Vocabulary **Review Vocabulary** Label the diagram with the correct terms. *(Lesson 8-1)*

leading coefficient

degree

$$12x^4 - x^3 + 2x + 5$$

Review Vocabulary Match the term with its definition by drawing a line to connect the two. *(Lesson 8-1)*

order of magnitude — the sum or difference of two monomials

trinomial — for a given quantity, the number rounded to the nearest power of 10

degree of a monomial — the sum of the exponents of all the variables of a monomial

binomial — the sum or difference of three monomials

Lesson 8-2 (continued)

Main Idea	Details
Polynomial Multiplied by Monomial	Follow the steps below to find $-2x^2(5x^2 - 3x + 1)$. 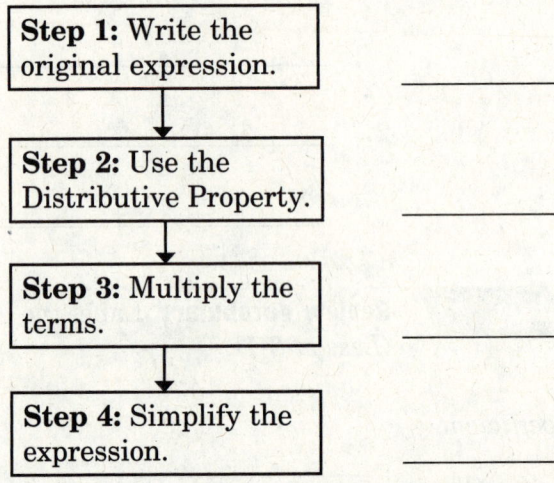
Solve Equations with Polynomial Expressions	Solve the equation below for m. Show your work. $m(m - 4) - m(m + 2) = -4m - 10$

$m = $ _____

Helping You Remember Use the equation $2x(x - 5) + 3x(x + 3) = 5x(x + 7) - 9$ to show how you would explain the process of solving equations with polynomial expressions to another algebra student.

8-3 Multiplying Polynomials

What You'll Learn

Scan the text under the *Now* heading. List two things you will learn about in the lesson.

1. _____

2. _____

Active Vocabulary

Review Vocabulary Write the definition next to each term. *(Lessons 8-1)*

order of magnitude ▶ _____

constant ▶ _____

New Vocabulary Fill in each blank with the correct term or phrase.

FOIL method ▶ To multiply two binomials using the FOIL method, find the sum of the products of **F** the _____, **O** the outer terms, **I** the inner terms, and **L** the _____.

quadratic expression ▶ A quadratic expression is an expression in one _____ with a degree of 2.

Chapter 8 — 135 — Glencoe Algebra 1

NAME _____ DATE _____ PERIOD _____

Lesson 8-3 (continued)

Main Idea	Details
Multiply Binomials	Use the FOIL method to find the product $(x + 8)(x - 5)$. 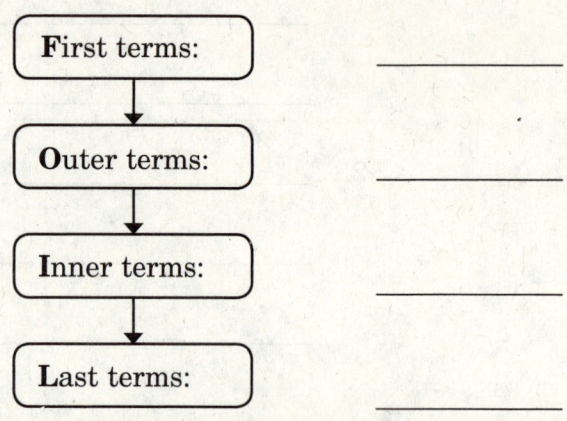 The product is _____.
Multiply Polynomials	Use the Distributive Property to find the product $(x + 1)(x^2 + x - 1)$. Show your work.

Helping You Remember

Think of a method for remembering all the product combinations used in the FOIL method for multiplying two binomials. Describe your method using words or a diagram.

_____

_____

_____

_____

8-4 Special Products

What You'll Learn

Scan the text in the lesson. Write two facts you learned about special products as you scanned the text.

1. _____

_____

_____

2. _____

_____

_____

Active Vocabulary

Review Vocabulary Match the term with its definition by drawing a line to connect the two. *(Lessons 7-4, 8-1, and 8-2)*

monomial   an expression in one variable with a degree of 2

quadratic expression   a monomial or the sum or difference of monomials, each called a *term*

polynomial   a form of a number that is written as $a \times 10^n$, where $1 \leq a < 10$ and n is an integer

scientific notation   a number, a variable, or the product of a number and one or more variables with nonnegative integer exponents

Review Vocabulary Label the diagram with the correct terms. *(Lesson 8-3)*

first

outer

inner $(x - 2)(x + 3) = x^2 + 3x - 2x - 6$

last

NAME _____ DATE _____ PERIOD _____

Lesson 8-4 (continued)

Main Idea	Details
Squares of Sums and Differences	Complete the tables to illustrate two special products.

Square of a Sum

Words	The square of $a + b$ is _____ _____ _____
Symbols	$(a + b)^2 =$
Example	$(n + 5)^2 =$

Square of a Difference

Words	The square of $a - b$ is _____ _____ _____
Symbols	$(a - b)^2 =$
Example	$(h - 7)^2 =$

Product of a Sum and a Difference

Find the product $(p - 2)(p + 2)$. Show your work.

Helping You Remember

Explain how the FOIL method can help you remember how many terms are in the special products studied in this lesson.

_____

_____

_____

_____

Chapter 8   138   Glencoe Algebra 1

NAME _____ DATE _____ PERIOD _____

8-5 Using the Distributive Property

What You'll Learn Scan the text under the *Now* heading. List two things you will learn about in the lesson.

1. _____
   _____
   _____

2. _____
   _____
   _____

Active Vocabulary **New Vocabulary** Write the definition next to each term.

factoring ▶ _____
_____
_____

factoring by grouping ▶ _____
_____
_____
_____

Zero Product Property ▶ _____
_____
_____
_____

Chapter 8

NAME _____ DATE _____ PERIOD _____

Lesson 8-5 (continued)

Main Idea	Details
Use the Distributive Property to Factor	Complete the following table illustrating when a polynomial can be factored by grouping.

Factoring by Grouping	
Words	A polynomial can be factored by grouping only if all of the following conditions exist. _____ _____ _____ _____ _____
Symbols	

Solve Equations by Factoring

Solve the following equation by factoring.

$4x^2 + 20x = 0$

(____)(____) = 0

_____ = 0 or _____ = 0

$x =$ _____ or $x =$ _____.

Helping You Remember

A good way to remember a mathematical concept is to explain it to somebody else. How would you help a classmate understand when it is possible to use the Zero Product Property to solve an equation? Give an example of an equation that can be solved using the Zero Product Property.

_____

_____

_____

Chapter 8   140   Glencoe Algebra 1

8-6 Solving $x^2 + bx + c = 0$

What You'll Learn

Skim the lesson. Predict two things that you expect to learn based on the headings and the Key Concept box.

1. _____

2. _____

Active Vocabulary

Review Vocabulary Match the term with its definition by drawing a line to connect the two. *(Lesson 8-2)*

Zero Product Property — four or more terms are put into groups and then factored

factoring — the largest number that is a factor of two numbers

factor by grouping — in order for a product to be equal to 0, at least one of the factors must be 0

New Vocabulary Write the correct term next to the definition.

_____ ▶ a type of equation that can be written in the standard form $ax^2 + bx + c = 0$, where $a \neq 0$

Chapter 8 — 141 — Glencoe Algebra 1

NAME _____ DATE _____ PERIOD _____

Lesson 8-6 (continued)

Main Idea	Details
Factor $x^2 + bx + c$	Factor $x^2 + 10x + 16$ by making an organized list of the factors of 16.

Factors of 16	Sum of factors

$x^2 + 10x + 16 = ($ _____ $)($ _____ $)$

Solve Equations by Factoring

Solve the quadratic equation $x^2 - 6x - 40 = 0$ by factoring.

$($ _____ $)($ _____ $) = 0$

_____ $= 0$ or _____ $= 0$

$x =$ _____ or $x =$ _____

Helping You Remember

If you are using the pattern $(x + m)(x + n)$ to factor a trinomial of the form $x^2 + bx + c$, how can you use your knowledge of multiplying integers to help you remember whether m and n are positive or negative?

_____

_____

_____

Chapter 8   142   Glencoe Algebra 1

8-7 Solving $ax^2 + bx + c = 0$

What You'll Learn

Scan the text in the lesson. Write two facts you learned about quadratic equations of the form $ax^2 + bx + c = 0$ as you scanned the text.

1. _____

_____

_____

_____

2. _____

_____

_____

_____

Active Vocabulary

New Vocabulary Write the definition next to the term.

prime polynomial ▶ _____

_____

_____

Vocabulary Link Recall the definition of a prime number. Describe how this definition relates to the definition of a prime polynomial.

_____

_____

_____

_____

Lesson 8-7 (continued)

Main Idea	Details
Factor $ax^2 + bx + c = 0$	Follow the steps below to factor the polynomial $2x^2 + 9x + 10$.

Step 1 Apply the pattern of factoring by grouping to write the desired form.

$2x^2 + 9x + 10 = 2x^2 + \underline{\hspace{1in}} + \underline{\hspace{1in}} + 10$

Step 2 Find two numbers that have a product of 2×10 or 20 and a sum of 9.

Factors of twenty   Sum of factors

Step 3 Use grouping to find the factors. Check your answer.

$2x^2 + 9x + 10 = (\underline{\hspace{0.7in}})(\underline{\hspace{0.7in}})$

Solve Equations by Factoring

Solve each equation. Check your solutions.

1. $2x^2 + 5x - 3 = 0$ 2. $3x^2 - 10x - 8 = 0$

$x = \underline{\hspace{0.7in}}$ $x = \underline{\hspace{0.7in}}$

Helping You Remember A good way to remember a mathematical procedure is to recite the steps of the procedure. What are the steps you would use to find the factors of a trinomial written in the form $ax^2 + bx + c = 0$?

NAME _____ DATE _____ PERIOD _____

8-8 Differences of Squares

What You'll Learn

Skim the Examples in the lesson. Predict two things you think you will learn about polynomials and quadratic equations that are differences of squares.

1. _____

_____

2. _____

_____

Active Vocabulary

New Vocabulary Circle each polynomial below that represents a difference of squares.

difference of squares ▶

$x^2 - 15$ $4b^2 - 49$

$3x^2 - 81$ $100n^2 - 1$

$16p^2 - 25$ $8r^2 - 12$

$256t^2 - 16$ $25h^2 - 4$

Vocabulary Link Describe how you can use the term *difference of squares* to recognize when a polynomial is of this form.

_____

_____

_____

Chapter 8   145   Glencoe Algebra 1

Lesson 8-8 (continued)

Main Idea	Details
Factor Differences of Squares	Model the process of factoring a polynomial that is a difference of squares by completing the following table.

Difference of Squares		
Symbols	$a^2 - b^2 =$ (_____) (_____)	
Examples	$x^2 - 16 =$ (_____) (_____)	
	$4y^2 - 1 =$ (_____) (_____)	
	$25 - 9g^2 =$ (_____) (_____)	

Solve Equations by Factoring

Solve $4n^2 - 25 = 0$ for n. Show your work.

Helping You Remember

A good way to remember a new mathematical concept is to explain it to a friend. Suppose a classmate is having difficulty remembering how to factor a difference of squares. How would you explain this concept to her?

_____

_____

_____

_____

NAME _____ DATE _____ PERIOD _____

8-9 Perfect Squares

What You'll Learn

Scan the lesson. List two headings you would use to make an outline of this lesson.

1. _____

   _____

2. _____

   _____

Active Vocabulary

Review Vocabulary Write the correct term next to each definition. *(Lessons 8-5, 8-6, and 8-7)*

_____ ▶ a monomial that is expressed as the product of prime numbers and variables, and no variable has an exponent greater than 1

_____ ▶ a process that involves writing a polynomial as the product of its factors

_____ ▶ a type of equation that can be written in the standard form $ax^2 + bx + c = 0$, where $a \neq 0$

_____ ▶ a polynomial that cannot be written as a product of two polynomials with integral coefficients

New Vocabulary Fill in the blank with the correct term or phrase.

perfect square trinomials ▶ Perfect square trinomials are trinomials that are the squares of _____.

Chapter 8   147   Glencoe Algebra 1

Lesson 8-9 (continued)

Main Idea	**Details**
Factor Perfect Square Trinomials	Model the process of factoring a polynomial that is a perfect square trinomial by completing the table.

Factoring Perfect Square Trinomials	
Symbols	$a^2 + 2ab + b^2 = (\underline{})^2$
	$a^2 - 2ab + b^2 = (\underline{})^2$
Examples	$x^2 + 8x + 16 = (\underline{})^2$
	$b^2 - 10b + 25 = (\underline{})^2$

Solve Equations with Perfect Squares

Use the Square Root Property to solve the equation $(x + 3)^2 = 100$. Check your solutions.

Helping You Remember Sometimes it is easier to remember a set of instructions if you can state them in a short sentence or phrase. Summarize the conditions that must be met in order for a trinomial to be factored as a perfect square trinomial.

NAME _____ DATE _____ PERIOD _____

Chapter 8: Quadratic Expressions and Equations

Tie It Together

Fill in the graphic organizer. Use examples from the chapter to add details if space permits.

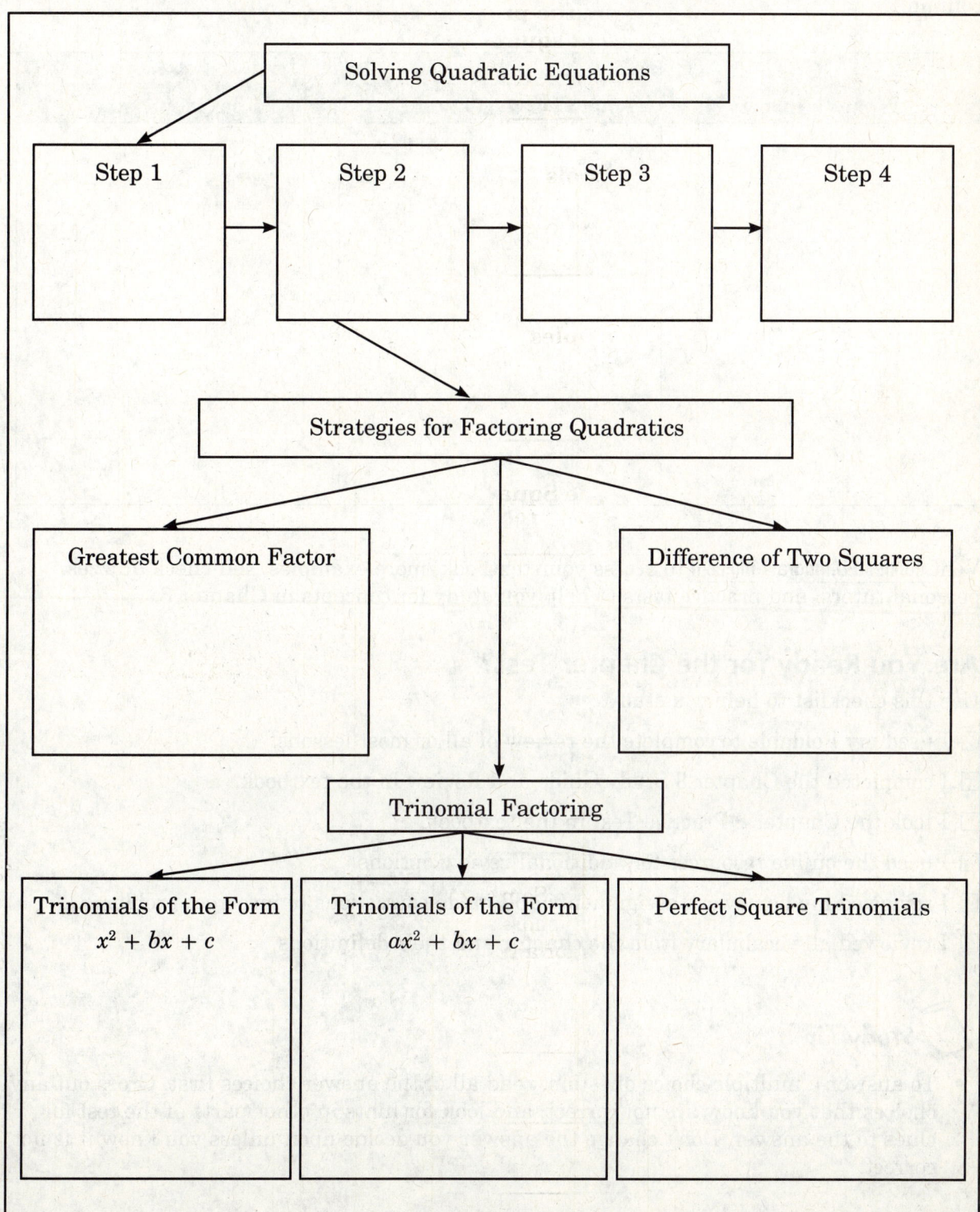

Chapter 8   149   Glencoe Algebra 1

NAME _____ DATE _____ PERIOD _____

Chapter 8: Quadratic Expressions and Equations

Before the Test

Review the ideas you listed in the table at the beginning of the chapter. Cross out any incorrect information in the first column. Then complete the table by filling in the third column.

K What I know...	W What I want to find out...	L What I learned...

Visit **connectED.mcgraw-hill.com** to access your textbook, more examples, self-check quizzes, personal tutors, and practice tests to help you study for concepts in Chapter 8.

Are You Ready for the Chapter Test?

Use this checklist to help you study.

☐ I used my Foldable to complete the review of all or most lessons.

☐ I completed the Chapter 8 Study Guide and Review in the textbook.

☐ I took the Chapter 8 Practice Test in the textbook.

☐ I used the online resources for additional review options.

☐ I reviewed my homework assignments and made corrections to incorrect problems.

☐ I reviewed all vocabulary from the chapter and their definitions.

 Study Tip

- To answer a multiple-choice question, read all of the answer choices first. Cross out any choices that you know are not correct, and look for hints in other parts of the test for clues to the answer. Don't change the answer you decide upon unless you know it is not correct.

Chapter 8   150   *Glencoe Algebra 1*

CHAPTER 9 — Quadratic Functions and Equations

Before You Read

Before you read the chapter, respond to these statements.
1. Write an **A** if you agree with the statement.
2. Write a **D** if you disagree with the statement.

Before You Read	Quadratic Functions and Equations
	• The graph of a quadratic function is a parabola.
	• When $a < 0$ in a quadratic function, the parabola opens up and has a minimum value.
	• The graph of $f(-x)$ flips the graph $f(x) = x^2$ across the x-axis.
	• Greatest integer functions can be written as piecewise linear functions.
	• Quadratic functions have constant first differences.

FOLDABLES Study Organizer Construct the Foldable as directed at the beginning of this chapter.

Note Taking Tips

- When you take notes, draw a visual (graph, diagram, picture, chart) that presents the information introduced in the lesson in a concise, easy-to-study format.
- In addition to writing important definitions in your notes, be sure to include your own examples of the concepts presented.

NAME _____ DATE _____ PERIOD _____

CHAPTER 9 Quadratic Functions and Equations

Key Points

Scan the pages in the chapter and write at least one specific fact concerning each lesson. For example, in the lesson on transformations of quadratic functions, one fact might be that a transformation changes the position or size of a figure. After completing the chapter, you can use this table to review for your chapter test.

Lesson	Fact
9-1 Graphing Quadratic Functions	
9-2 Solving Quadratic Equations by Graphing	
9-3 Transformations of Quadratic Functions	
9-4 Solving Quadratic Equations by Completing the Square	
9-5 Solving Quadratic Equations by Using the Quadratic Formula	
9-6 Analyzing Functions with Successive Differences	
9-7 Special Functions	

Chapter 9   Glencoe Algebra 1

9-1 Graphing Quadratic Functions

What You'll Learn Skim the lesson. Predict two things that you expect to learn based on the headings and the Key Concept box.

1. _____

2. _____

Active Vocabulary **New Vocabulary** Write the correct term next to each definition.

_____ ▶ a function with a graph that is not a straight line

_____ ▶ a nonlinear function that can be written in the form $f(x) = ax^2 + bx + c$, where $a \neq 0$

_____ ▶ the form of a quadratic function when it is written as $f(x) = ax^2 + bx + c$

_____ ▶ the shape of the graph of a quadratic function

_____ ▶ the central line about which a parabola is symmetric

_____ ▶ the point of intersection between a parabola and its axis of symmetry

_____ ▶ the lowest point on a parabola

_____ ▶ the highest point on a parabola

NAME _____ DATE _____ PERIOD _____

Lesson 9-1 (continued)

Main Idea	Details
Characteristics of Quadratic Functions	Fill in the boxes with the correct terms. 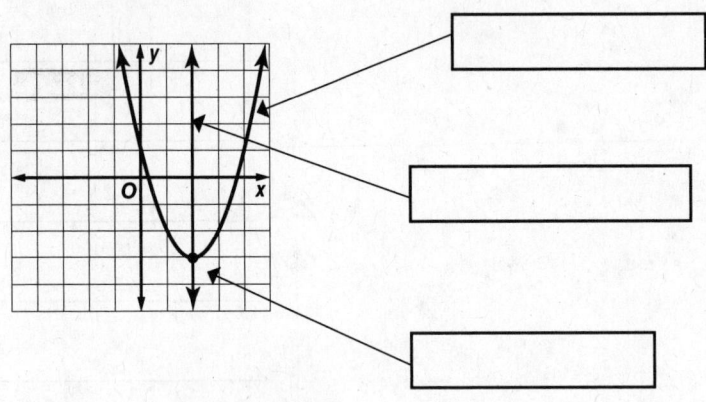
Graph Quadratic Functions	Graph the quadratic function $f(x) = x^2 + 3x + 2$ on the coordinate grid below. 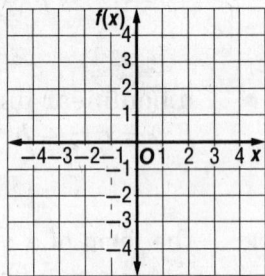

Helping You Remember Look up the word *vertex* in a dictionary. You will find that it comes from the Latin word *vertere*, which means to turn. How can you use the idea of "to turn" to remember the vertex of a parabola?

Chapter 9   154   Glencoe Algebra 1

9-2 Solving Quadratic Equations by Graphing

What You'll Learn

Scan the lesson. List two headings you would use to make an outline of this lesson.

1. _____

2. _____

Active Vocabulary

Review Vocabulary Match each term with its definition by drawing a line to connect the two. *(Lesson 9-1)*

nonlinear function   the shape of the graph of a quadratic function

minimum   the central line about which a parabola is symmetric

parabola   a function with a graph that is not a straight line

axis of symmetry   the point of intersection between a parabola and its axis of symmetry

vertex   the lowest point on a parabola

New Vocabulary Write the definition next to the term.

double root ▶ _____

Lesson 9-2 (continued)

Main Idea	Details
Solve by Graphing	Complete the following table illustrating the number and nature of the solutions of a quadratic equation.

Solutions of Quadratic Equations	
Number of real solutions	Relationship with the *x*-axis
0	
1	
2	

Estimate Solutions

Solve the quadratic equation below by graphing. If integral roots cannot be found, estimate the roots to the nearest tenth.

$x^2 + 3x - 2 = 0$

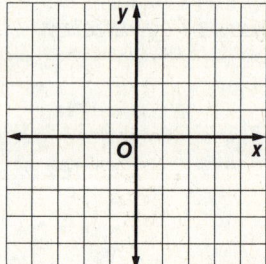

$x = $ _____

Helping You Remember

Describe how you can remember that the word *zero* is used when you are talking about functions, but the word *root* is used when you are talking about equations.

_____

_____

9-3 Transformations of Quadratic Functions

What You'll Learn

Skim the lesson. Write two things you already know about transformations of quadratic functions.

1. _____

_____

_____

_____

2. _____

_____

_____

_____

Active Vocabulary

New Vocabulary Fill in each blank with the correct term or phrase.

dilation ▶ A dilation makes the graph narrower or wider than the _____.

reflection ▶ A reflection flips a figure over a _____.

transformation ▶ A transformation changes the _____ or _____ of a figure.

translation ▶ A translation moves a figure _____, down, or _____.

Lesson 9-3 (continued)

Main Idea	Details
Translations	Describe how the graph of each function is related to the graph of $f(x) = x^2$. 1. $f(x) = x^2 - 6$ 2. $f(x) = x^2 + \dfrac{1}{2}$ _____ _____ _____ _____
Dilations and Reflections	Complete the table below by naming and describing each transformation of $f(x)$.

Dilations and Reflections

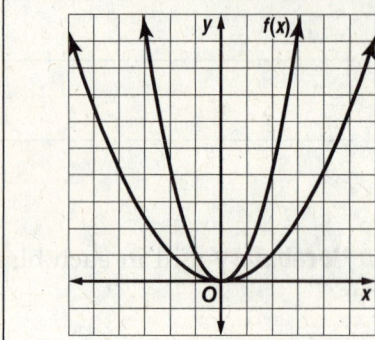

Transformation: _____

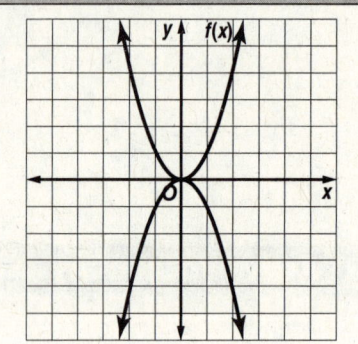

Transformation: _____

Helping You Remember

A good way to remember mathematical terms is to relate them to a term you already know. Translations are often called *slides*, and reflections are often called *flips*. Explain how these terms accurately describe the corresponding transformations of parabolas.

_____

_____

_____

9-4 Solving Quadratic Equations by Completing the Square

What You'll Learn

Scan the text in the lesson. Write two facts you learned about solving quadratic equations by completing the square as you scanned the text.

1. _____

_____

_____

_____

2. _____

_____

_____

_____

Active Vocabulary

Review Vocabulary Label each diagram with the correct term to describe the transformation. (*Lesson 9-3*)

dilation ▶

translation ▶

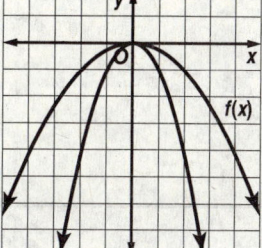   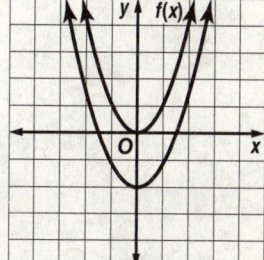

New Vocabulary Fill in the blank with the correct term or phrase.

completing the square ▶ Any quadratic expression in the form $x^2 + bx$ can be made into a _____ trinomial by using a method called completing the square.

NAME _____ DATE _____ PERIOD _____

Lesson 9-4 (continued)

Main Idea	Details
Complete the Square	Complete the following table to show the steps that you must follow to complete the square.

Completing the Square	
Words	To complete the square for any quadratic expression of the form $x^2 + bx$, follow the steps below. Step 1 _____ _____ Step 2 _____ Step 3 _____
Symbols	$x^2 + bx +$

Solve Equations by Completing the Square

Solve $x^2 + 6x = 27$ by completing the square. Show your work.

Helping You Remember How is completing the square related to the method you use to determine whether a trinomial is a perfect square trinomial?

_____

_____

_____

Chapter 9   160   Glencoe Algebra 1

9-5 Solving Quadratic Equations by Using the Quadratic Formula

What You'll Learn

Skim the Examples in the lesson. Predict two things you think you will learn about solving quadratic equations by using the quadratic formula.

1. _____

   _____

   _____

2. _____

   _____

   _____

   _____

Active Vocabulary

Review Vocabulary Write the correct term next to each definition. (*Lesson 9-1*)

_____ ▶ the shape of the graph of a quadratic function

_____ ▶ the central line about which a parabola is symmetric

_____ ▶ the highest point on a parabola

New Vocabulary Write the correct term next to each definition.

_____ ▶ the formula that gives the solutions to the general quadratic equation, $ax^2 + bx + c = 0$, as $x = \dfrac{-b \pm \sqrt{b^2 - 4ac}}{2a}$

_____ ▶ the expression under the radical sign in the quadratic formula, $b^2 - 4ac$

NAME _____ DATE _____ PERIOD _____

Lesson 9-5 (continued)

Main Idea	Details
Quadratic Formula	Solve the equation $x^2 + 5x + 3 = 0$ by using the Quadratic Formula. Determine the exact solutions. Show your work.

| The Discriminant | Complete the following table to show the relationship between the discriminant of a quadratic equation and its solutions and graph. |

The Discriminant		
Value	Number of real solutions	Relationship between graph and the *x*-axis
$b^2 - 4ac > 0$		
$b^2 - 4ac = 0$		
$b^2 - 4ac < 0$		

Helping You Remember To help remember the methods for solving a quadratic equation, explain how you would choose the best method for solving a form of the quadratic equation $ax^2 + bx + c = 0$.

_____

_____

_____

Chapter 9 162 Glencoe Algebra 1

9-6 Analyzing Functions with Successive Differences and Ratios

What You'll Learn

Scan the text under the *Now* heading. List two things you will learn about in the lesson.

1. _____

_____

_____

2. _____

_____

_____

_____

Active Vocabulary

Review Vocabulary Write the correct term next to each definition. *(Lessons 7-5, 7-6, 9-1, 9-3, 9-4, and 9-5)*

_____ ▶ a function with a graph that is not a straight line

_____ ▶ the form of a quadratic function when it is written as $f(x) = ax^2 + bx + c$

_____ ▶ a transformation that makes a function wider or narrower than the parent function

_____ ▶ the expression under the radical sign in the quadratic formula, $b^2 - 4ac$

_____ ▶ a function of the form $y = ab^x$, where $a \neq 0$, $b > 0$, and $b \neq 1$

_____ ▶ the ratio of two consecutive terms in a geometric sequence

NAME _____ DATE _____ PERIOD _____

Lesson 9-6 (continued)

Main Idea	Details
Identify Functions	Complete the table below by writing the general form of each function and sketching a sample graph.

Linear, Quadratic, and Exponential Functions		
Linear Function	**Quadratic Function**	**Exponential Function**
$y = mx + b$	$y = ax^2 + bx + c$	$y = ab^x$

Write Equations

Determine which model best describes the data in the table. Then write an equation for the function that models the data.

x	−5	−4	−3	−2	−1
y	160	80	40	20	10

Helping You Remember

A good way to remember a mathematical concept is to explain it in your own words. Explain how you can determine the type of a function simply by looking at its graph.

_____

_____

_____

9-7 Special Functions

What You'll Learn

Scan the lesson. List two headings you would use to make an outline of this lesson.

1. _____

2. _____

Active Vocabulary

Review Vocabulary Graph each on a number line. *(Lesson 2-5)*

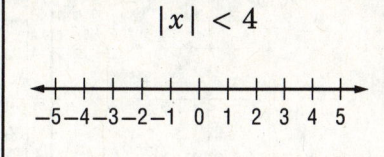

New Vocabulary Match the term with its definition by drawing a line to connect the two.

piecewise-defined function — a function whose graph consists of disjointed line segments

step function — a function when given x, returns the greatest integer less than or equal to x

piecewise linear function — a function written using two or more expressions

absolute value function — a function which contains an algebraic expression within absolute value symbols

greatest integer function — a function written using one expression which results in a graph that consists of multiple lines

NAME _____ DATE _____ PERIOD _____

Lesson 9-7 (continued)

Main Idea	Details
Step Functions	Evaluate each expression. 1. $[\![8.7]\!]$ 　　2. $[\![-8.2]\!] + [\![16.2]\!]$ 3. $[\![12.1]\!] + 8$ 　　4. $[\![18.9 + 12.6]\!]$ Provide either the graph or the function notation for each *piecewise-defined function*. Identify the domain and range for each.

Absolute Value Functions

Function	Graph	Domain
$f(x) = \begin{cases} 3x - 5 & \text{if } x < 3 \\ -\dfrac{2}{3}x - 4 & \text{if } x \geq 3 \end{cases}$		
		Range
		Domain
		Range

Helping You Remember Explain how you can use a number line to find the value of the greatest integer function for any real number.

_____

_____

Chapter 9: Quadratic Functions and Equations

Tie It Together

Fill in each graphic organizer paying attention to the depicted relationships between the organizers. Add details for each organizer.

Quadratic Equations – Solution Methods

Method	Graphing	Factoring	Square Root Property	Completing the Square	Quadratic Formula
Description					

Graphing a Quadratic
- Vertex
- Axis of Symmetry
- Shape

Possible Discriminant Values and Solution Details

Using Equations to Model Data . . . Given a Table of Values

Linear Data
Form:
Detecting:

Quadratic Data
Form:
Detecting:

Exponential Data
Form:
Detecting:

NAME _____ DATE _____ PERIOD _____

 # Quadratic Functions and Equations

Before the Test

Now that you have read and worked through the chapter, think about what you have learned and complete the table below. Compare your previous answers with these.

1. Write an **A** if you agree with the statement.
2. Write a **D** if you disagree with the statement.

Quadratic Functions and Equations	After You Read
• The graph of a quadratic function is a parabola.	
• When $a < 0$ in a quadratic function, the parabola opens up and has a minimum value.	
• The graph of $f(-x)$ flips the graph $f(x) = x^2$ across the x-axis.	
• Greatest integer functions can be written as piecewise linear functions.	
• Quadratic functions have constant first differences.	

Visit **connectED.mcgraw-hill.com** to access your textbook, more examples, self-check quizzes, personal tutors, and practice tests to help you study for concepts in Chapter 9.

Are You Ready for the Chapter Test?

Use this checklist to help you study.

☐ I used my Foldable to complete the review of all or most lessons.

☐ I completed the Chapter 9 Study Guide and Review in the textbook.

☐ I took the Chapter 9 Practice Test in the textbook.

☐ I used the online resources for additional review options.

☐ I reviewed my homework assignments and made corrections to incorrect problems.

☐ I reviewed all vocabulary from the chapter and their definitions.

• If possible, rewrite your notes. Not only can you make them clearer and neater, rewriting them will help you remember the information.

Chapter 10 Radical Functions and Geometry

Before You Read

Before you read the chapter, respond to these statements.
1. Write an **A** if you agree with the statement.
2. Write a **D** if you disagree with the statement.

Before You Read	Radical Functions and Geometry
	• The graph of a square root function includes both positive and negative values.
	• $tan\ A = \dfrac{opposite}{adjacent}$
	• The product of two conjugates is a rational number.
	• In a Pythagorean triple, two or three numbers can be equal.
	• Rationalizing the denominator is the process that eliminates a radical from the denominator of a fraction.

FOLDABLES Study Organizer Construct the Foldable as directed at the beginning of this chapter.

 Note Taking Tips

- Take notes in such a manner that someone who did not understand the topic will understand after reading what you have written.

- When you take notes, write a summary of the lesson, or write in your own words what the lesson was about.

NAME _____ DATE _____ PERIOD _____

CHAPTER 10 Radical Functions and Geometry

Key Points

Scan the pages in the chapter and write at least one specific fact concerning each lesson. For example, in the lesson on the Pythagorean Theorem, one fact might be that in a right triangle, the side opposite the right angle is the hypotenuse. After completing the chapter, you can use this table to review for your chapter test.

Lesson	Fact
10-1 Square Root Functions	
10-2 Simplifying Radical Expressions	
10-3 Operations with Radical Expressions	
10-4 Radical Equations	
10-5 The Pythagorean Theorem	
10-6 Trigonometric Ratios	

Chapter 10   170   Glencoe Algebra 1

NAME _____ DATE _____ PERIOD _____

10-1 Square Root Functions

What You'll Learn Skim the lesson. Write two things you already know about square root functions.

1. _____

2. _____

Active Vocabulary

New Vocabulary Match the term with its definition by drawing a line to connect the two.

radicand   a function that contains a variable under a radical sign

radical function   a function that contains the square root of a variable

square root function   the expression under the radical sign

Vocabulary Link Recall that the square root of a negative number is not defined to be a real number. Explain what effect this has on the domain of a square root function.

_____

NAME _____ DATE _____ PERIOD _____

Lesson 10-1 (continued)

Main Idea	Details

Dilations of Radical Functions

Graph $f(x) = \frac{1}{2}\sqrt{x}$. State the domain and range.

Step 1: Make a table of function values.

x	0	1	4	9	16
f(x)					

Step 2: Plot the points on a coordinate grid.

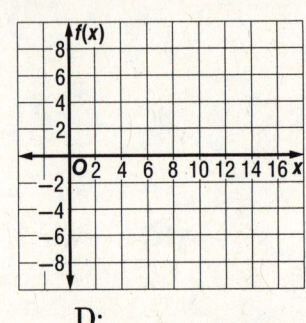

Step 3: Connect the points with a smooth curve.

D: _____
R: _____

Reflections and Translations of Radical Functions

The velocity of an object dropped from a height of h meters is given by the function $v = \sqrt{2gh}$, where g is the constant, 9.8 meters per second squared. What is the velocity of an object when it hits the ground if it is dropped from a height of 100 meters? Show your work and round your answer to the nearest tenth.

$v \approx$ _____ m/s

Helping You Remember Suppose a classmate is having difficulty remembering how to graph a square root function. What advice would you give him about how to select suitable domain values?

_____
_____
_____
_____

Chapter 10   172   Glencoe Algebra 1

NAME _____ DATE _____ PERIOD _____

10-2 Simplifying Radical Expressions

What You'll Learn

Scan the text under the *Now* heading. List two things you will learn about in the lesson.

1. _____

_____

_____

_____

2. _____

_____

_____

_____

Active Vocabulary

Review Vocabulary Write the definition next to the term. *(Lesson 10-1)*

radicand ▶ _____

_____

New Vocabulary Fill in each blank with the correct term or phrase.

conjugate ▶ Binomials of the form $a\sqrt{b} + c\sqrt{d}$ and _____ are called conjugates.

radical expression ▶ A radical expression contains a _____, such as a square root.

rationalizing the denominator ▶ Rationalizing the denominator of a fraction with a radical eliminates all _____ from the _____.

Chapter 10   173   Glencoe Algebra 1

Lesson 10-2 *(continued)*

Main Idea	Details
Product Property of Square Roots	Complete the following table to illustrate the Product Property of Square Roots.

Product Property of Square Roots	
Words	For any nonnegative real numbers a and b, the square roots of ab is equal _____ _____ _____
Symbols	$\sqrt{ab} = $ _____
Example	$\sqrt{16 \cdot 25} = $ _____

Quotient Property of Square Roots

Simplify the expression $\dfrac{2}{4 + \sqrt{5}}$. Show your work.

Helping You Remember

What should you remember to check for when you want to determine if a radical expression is in simplest form?

_____

_____

_____

Chapter 10 — Glencoe Algebra 1

10-3 Operations with Radical Expressions

What You'll Learn

Skim the Examples in this lesson. Predict two things you think you will learn about operations with radical expressions.

1. _____

   _____

   _____

   _____

2. _____

   _____

   _____

   _____

Active Vocabulary

Review Vocabulary Write the correct term next to each definition. *(Lessons 10-1 and 10-2)*

_____ ▶ a function that contains the square root of a variable

_____ ▶ an expression that contains a radical, such as a square root

_____ ▶ the expression under the radical sign

_____ ▶ binomials of the form $a\sqrt{b} + c\sqrt{d}$ and $a\sqrt{b} - c\sqrt{d}$

_____ ▶ a function that contains a variable under a radical sign

_____ ▶ a process that eliminates all radicals from the denominator of a fraction

Chapter 10   175   Glencoe Algebra 1

NAME _____ DATE _____ PERIOD _____

Lesson 10-3 (continued)

Main Idea	**Details**
Add or Subtract Radical Expressions	Simplify each expression in the table illustrating how adding and subtracting radical expressions is similar to adding and subtracting monomials.

Monomials	Radical Expressions
$6b + 3b = $ _____ $= $ _____	$6\sqrt{2} + 3\sqrt{2} = $ _____ $= $ _____
$12m - 5m = $ _____ $= $ _____	$12\sqrt{3} - 5\sqrt{3} = $ _____ $= $ _____

Multiply Radical Expressions

Simplify each expression. Show your work.

1. $6\sqrt{5}\,(2\sqrt{5})$

2. $2\sqrt{3}\,(6\sqrt{7} - \sqrt{7})$

_____ _____

Helping You Remember How can you use what you know about adding and subtracting monomials to help you remember how to add and subtract radical expressions?

_____

_____

_____

_____

10-4 Radical Equations

What You'll Learn

Scan the lesson. List two headings you would use to make an outline of this lesson.

1. _____

_____

2. _____

_____

Active Vocabulary

New Vocabulary Fill in each blank with the correct term or phrase.

radical equations ▶ Equations that contain variables in the _____, like $h = 1.34\sqrt{\ell}$, are called radical equations.

extraneous solutions ▶ Squaring each side of an equation sometimes produces a _____ that is not a solution of the original _____. These are called extraneous solutions.

Vocabulary Link Look up the definition of *extraneous* in a dictionary. Describe how this definition applies to the concept of extraneous solutions.

_____

_____

_____

_____

_____

Lesson 10-4 (continued)

Main Idea	Details
Radical Equations	Solve the equation $\sqrt{n+4} + 2 = 5$. Show your work.
Extraneous Solutions	Follow the steps below to solve the equation $x = \sqrt{x+3} - 1$. Check your solutions.

Step 1: Isolate the radical on one side.

Step 2: Square each side to eliminiate the radical.

Step 3: Solve. Check for extraneous solutions.

solution(s): _____   extraneous solution(s): _____

Helping You Remember

Acronyms can be a useful tool for remembering the steps in a mathematical process. For example, the acronym FOIL reminds you to multiply the First, Outer, Inner, and Last terms when multiplying two binomials. How can you use the letters ISC to remember the three steps in solving a radical equation?

10-5 The Pythagorean Theorem

What You'll Learn

Skim the lesson. Predict two things that you expect to learn based on the headings and the Key Concept box.

1. _____

2. _____

Active Vocabulary

Review Vocabulary Write the definition next to each term. *(Lessons 10-1 and 10-4)*

radical function ▶ _____

radical equation ▶ _____

New Vocabulary Match the term with its definition by drawing a line to connect the two.

converse — the two shorter sides of a right triangle

hypotenuse — the result when the hypothesis and conclusion of an if-then statement are exchanged

legs — a group of three whole numbers that satisfy the equation $c^2 = a^2 + b^2$, where c is the greatest number

Pythagorean Triple — the side opposite the right angle in a right triangle

Lesson 10-5 (continued)

Main Idea	Details
The Pythagorean Theorem	Label the sides of the right triangle shown using the terms *leg* and *hypotenuse*. Then write an equation to demonstrate the Pythagorean Theorem. 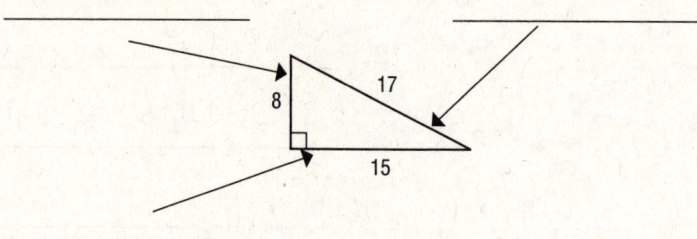
Right Triangles	Circle the sets of numbers below that represent Pythagorean triples. 1. 5, 12, 13 2. 18, 24, 30 3. 16, 32, 34 4. 30, 40, 50 5. 9, 40, 41 6. 8, 12, 20

Helping You Remember

Think of a word or phrase that you can associate with the Pythagorean Theorem to help you remember the equation $c^2 = a^2 + b^2$.

_____

_____

_____

_____

10-6 Trigonometric Ratios

What You'll Learn

Scan the lesson. List two headings you would use to make an outline of this lesson.

1. _____

_____

2. _____

_____

Active Vocabulary

New Vocabulary Write the correct term next to each definition.

_____ ▶ the ratio of the opposite leg to the hypotenuse of a right triangle

_____ ▶ a function that has a rule given by a trigonometric ratio

_____ ▶ the measure of ∠A if sin A is known

_____ ▶ the study of triangle measurement

_____ ▶ the measure of ∠A if tan A is known

_____ ▶ finding all unknown sides and angles of a right triangle

_____ ▶ the ratio of the adjacent leg to the hypotenuse of a right triangle

_____ ▶ the measure of ∠A if cos A is known

_____ ▶ a ratio of the lengths of two sides of a right triangle

_____ ▶ the ratio of the opposite leg to the adjacent leg of a right triangle

NAME _____ DATE _____ PERIOD _____

Lesson 10-6 (continued)

Main Idea	Details
Trigonometric Ratios	Complete the chart to show the trigonometric ratios for angles R and S. 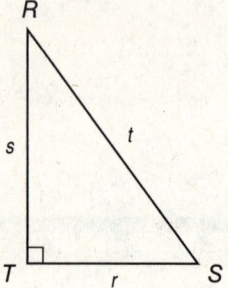 $\sin R =$ _____ $\sin S =$ _____ $\cos R =$ _____ $\cos S =$ _____ $\tan R =$ _____ $\tan S =$ _____
Use Trigonometric Ratios	Use a calculator to find the measure of $\angle M$ to the nearest tenth. 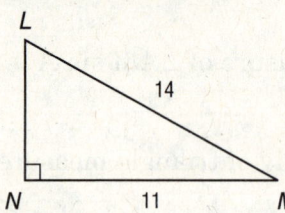 $m\angle M \approx$ _____

Helping You Remember

How can the *co* in *cosine* help you to remember the relationship between the sines and the cosines of the two acute angles of a right triangle?

_____

_____

_____

Chapter 10   182   Glencoe Algebra 1

NAME _____ DATE _____ PERIOD _____

CHAPTER 10: Radical Functions and Geometry

Tie It Together

Fill in details in the organizer.

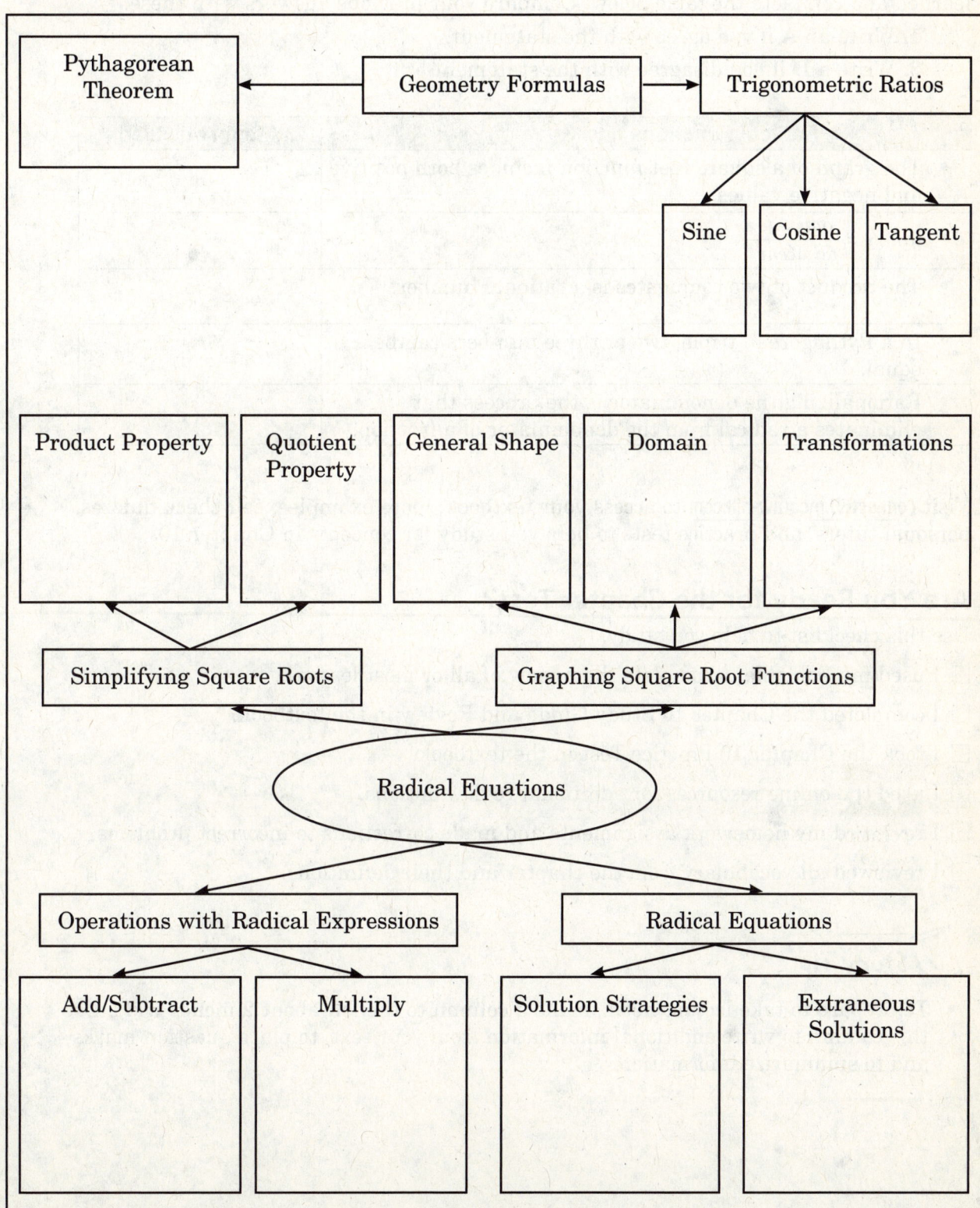

Chapter 10 — 183 — Glencoe Algebra 1

NAME _____ DATE _____ PERIOD _____

Chapter 10 Radical Functions and Geometry

Before the Test

Now that you have read and worked through the chapter, think about what you have learned and complete the table below. Compare your previous answers with these.

1. Write an **A** if you agree with the statement.
2. Write a **D** if you disagree with the statement.

Radical Functions and Geometry	After You Read
• The graph of a square root function includes both positive and negative values.	
• $\tan A = \dfrac{opposite}{adjacent}$	
• The product of two conjugates is a rational number.	
• In a Pythagorean triple, two or three numbers can be equal.	
• Rationalizing the denominator is the process that eliminates a radical from the denominator of a fraction.	

Visit connectED.mcgraw-hill.com to access your textbook, more examples, self-check quizzes, personal tutors, and practice tests to help you study for concepts in Chapter 10.

Are You Ready for the Chapter Test?

Use this checklist to help you study.

☐ I used my Foldable to complete the review of all or most lessons.

☐ I completed the Chapter 10 Study Guide and Review in the textbook.

☐ I took the Chapter 10 Practice Test in the textbook.

☐ I used the online resources for additional review options.

☐ I reviewed my homework assignments and made corrections to incorrect problems.

☐ I reviewed all vocabulary from the chapter and their definitions.

Study Tip

- To prepare to take lecture notes, make a column to the left about 2 inches wide. Use this column to write additional information from your text, to place question marks, and to summarize information.

NAME _____ DATE _____ PERIOD _____

Rational Functions and Equations

Before You Read

Before you read the chapter, think about what you know about rational functions and equations. List three things you already know about them in the first column. Then list three things you would like to learn about them in the second column.

K What I know...	W What I want to find out...

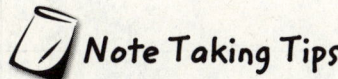 Construct the Foldable as directed at the beginning of this chapter.

Note Taking Tips

- **When taking notes, place a question mark next to anything you do not understand.**
  Then be sure to ask questions before any quizzes or tests.

- **When you take notes in geometry, be sure to make comparisons among the different formulas and concepts.**
  For example, how are pyramids and cones similar? Different? This will help you learn the material.

Chapter 11 185 Glencoe Algebra 1

NAME _____ DATE _____ PERIOD _____

Rational Functions and Equations

Key Points

Scan the pages in the chapter and write at least one specific fact concerning each lesson. For example, in the lesson on rational functions, one fact might be that a line that the graph of a function approaches is called an asymptote. After completing the chapter, you can use this table to review for your chapter test.

Lesson	Fact
11-1 Inverse Variation	
11-2 Rational Functions	
11-3 Simplifying Rational Expressions	
11-4 Multiplying and Dividing Rational Expressions	
11-5 Dividing Polynomials	
11-6 Adding and Subtracting Rational Expressions	
11-7 Mixed Expressions and Complex Fractions	
11-8 Rational Equations	

Chapter 11   186   Glencoe Algebra 1

11-1 Inverse Variation

What You'll Learn

Skim the Examples in the lesson. Predict two things you think you will learn about inverse variation.

1. _____

_____

_____

2. _____

_____

_____

Active Vocabulary

New Vocabulary Fill in each blank with the correct term or phrase.

inverse variation ▶ A relationship between x and y that can be represented by the equation _____ or _____ is an inverse variation.

product rule ▶ The equation _____ is called the product rule for inverse variations.

Vocabulary Link Look up the word *inverse* in a dictionary. Explain how the definition applies to an inverse variation equation.

_____

_____

_____

_____

NAME _____ DATE _____ PERIOD _____

Lesson 11-1 (continued)

Main Idea	Details
Identify and Use Inverse Variations	Determine whether the data in the table represent an *inverse* or a *direct* variation. Explain.

x	4	8	12	16
y	36	18	12	9

_____

_____

Graph Inverse Variations

Sketch a sample direct and inverse variation on the coordinate grids below, and complete the table.

Direct Variation	Inverse Variation
$y = $ _____	$y = $ _____
y varies _____ as x.	y varies _____ as x.
The ratio _____ is constant.	The product _____ is constant.

Helping You Remember

A good way to help you remember mathematical concepts is to state them in your own words. To remember how to set up a proportion to solve a problem involving inverse variation, write a sentence describing the form the proportion should have.

_____

_____

Chapter 11   Glencoe Algebra 1

NAME _____ DATE _____ PERIOD _____

11-2 Rational Functions

What You'll Learn Skim the lesson. Write two things you already know about rational functions.

1. _____
   _____
   _____
   _____

2. _____
   _____

Active Vocabulary **New Vocabulary** Write the definition next to each term.

asymptote ▶ _____
   _____
   _____

excluded value ▶ _____
   _____
   _____

rational function ▶ _____
   _____
   _____

Vocabulary Link Explain why the term *rational* is used to describe a rational function.

_____
_____

Lesson 11-2 (continued)

Main Idea	Details
Identify Excluded Values	State the excluded value for each function. 1. $y = \dfrac{7}{x+6}$ 2. $y = \dfrac{-5}{4x-20}$ _____ _____ 3. $y = \dfrac{4}{3x+21}$ 4. $y = \dfrac{-1}{-2x-8}$ _____ _____
Identify and Use Asymptotes	The rational function $y = \dfrac{1}{x-3} + 2$ is graphed below. Identify the asymptotes. 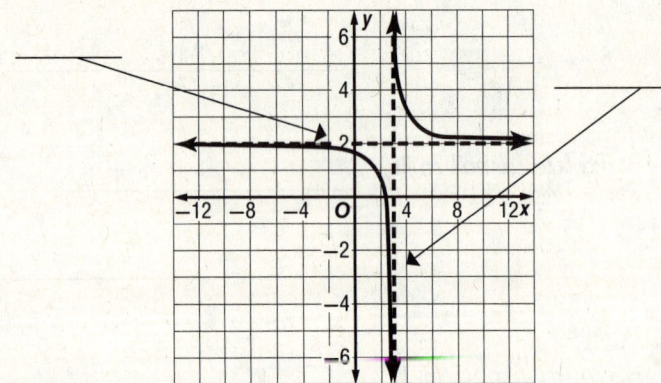

Helping You Remember

A good way to remember a mathematical concept is to explain it to someone else. Suppose a classmate is having difficulty finding the excluded values of a rational function. How would you explain the process?

_____

_____

_____

Chapter 11   Glencoe Algebra 1

11-3 Simplifying Rational Expressions

What You'll Learn

Scan the text under the *Now* heading. List two things you will learn about in the lesson.

1. _____

   _____

   _____

2. _____

   _____

   _____

Active Vocabulary

Review Vocabulary Match the term with its definition by drawing a line to connect the two. *(Lessons 11-1 and 11-2)*

rational function — a line that the graph of a rational function approaches, but never crosses or touches

asymptote — a relationship between x and y that can be represented by the equation $y = \frac{k}{x}$ or $xy = k$

inverse variation — a value that results in zero in the denominator of a rational function

excluded value — a function that can be described by the equation $y = \frac{p}{q}$, where p and q are polynomials and $q \neq 0$

New Vocabulary Write the correct term next to the definition.

_____ ▶ an algebraic fraction whose numerator and denominator are polynomials, such as $\frac{2x - 1}{x^2 + 5x + 9}$

Lesson 11-3 (continued)

Main Idea	Details
Identify Excluded Values	Follow the steps below to find the excluded values for the rational expression $\dfrac{-12}{b^2 - 49}$.
	Step 1: Set the denominator equal to zero. _____
	Step 2: Factor the expression. _____
	Step 3: Solve for the excluded values. _____
Simplify Expressions	Simplify the expression $\dfrac{x^2 + 3x - 18}{x - 3}$. State the excluded value(s). _____

Helping You Remember   Mathematical concepts are often built on ideas that you learned in previous classes. Explain how you can use what you know about simplifying fractions for rational numbers to remember how to simplify rational expressions.

_____
_____
_____

11-4 Multiplying and Dividing Rational Expressions

What You'll Learn

Scan the text in the lesson. Write two facts you learned about multiplying and dividing rational expressions as you scanned the text.

1. _____

   _____

   _____

2. _____

   _____

   _____

   _____

Active Vocabulary

Review Vocabulary Label the diagram with the correct terms. *(Lesson 11-2)*

asymptotes ▶

rational function ▶

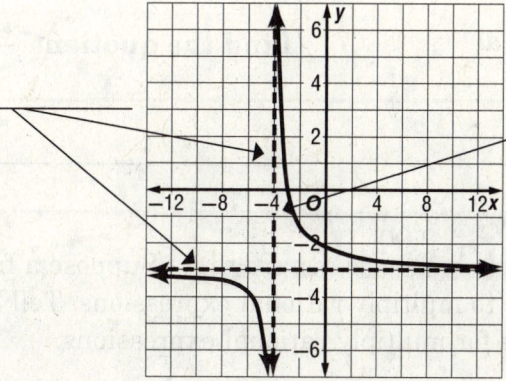

Review Vocabulary Write the definition next to the term. *(Lesson 11-1)*

product rule (for inverse variations) ▶ _____

_____

_____

Chapter 11   193   Glencoe Algebra 1

NAME _____ DATE _____ PERIOD _____

Lesson 11-4 (continued)

Main Idea	Details
Multiply Rational Expressions	Follow the steps below to simplify $\dfrac{1}{n^2 - 25} \cdot \dfrac{n^2 + 7n + 10}{n + 2}$.

Step 1: Factor the numerators and denominators.

_____

Step 2: Cross out common factors.

_____

Step 3: Simplify the expression.

_____

Divide Rational Expressions

Find the quotient $\dfrac{6x - 12}{x^2} \div (x - 2)$.

_____

Helping You Remember

Suppose a friend was absent when the class was studying how to multiply rational expressions. Tell how you can explain to your friend the procedure for multiply rational expressions.

_____

_____

_____

_____

NAME _____ DATE _____ PERIOD _____

11-5 Dividing Polynomials

What You'll Learn

Skim the lesson. Predict two things that you expect to learn based on the headings and figures in the lesson.

1. _____

   _____

2. _____

   _____

Active Vocabulary

Review Vocabulary Write the definition next to each term. (Lessons 11-1, 11-2, and 11-3)

asymptote ▶ _____

_____

excluded value ▶ _____

_____

rational expression ▶ _____

_____

inverse variation ▶ _____

_____

NAME _____ DATE _____ PERIOD _____

Lesson 11-5 (continued)

Main Idea	Details
Divide Polynomials by Monomials	Find each quotient. Show your work. 1. $(5x^2 - 10x) \div 5x$ 2. $(2n^2 - 9n + 4) \div 2n$ 3. $(12m^2 + 9m) \div -3m$ 4. $(4y^2 + 18y - 6) \div 6y$
Divide Polynomials by Binomials	Find $(h^2 + 6h - 40) \div (h - 4)$ by using long division. Show your work.

Helping You Remember

There are several methods you can use to divide polynomials by binomials. If you want to remember one method that you can always use to divide a polynomial by a binomial, which method should you select? Explain.

_____

_____

_____

_____

11-6 Adding and Subtracting Rational Expressions

What You'll Learn

Scan the lesson. List two headings you would use to make an outline of this lesson.

1. _____

_____

2. _____

_____

Active Vocabulary

Review Vocabulary Write the correct term next to each definition. *(Lessons 1-1, 11-2, and 11-3)*

_____ ▶ an algebraic fraction whose numerator and denominator are polynomials

_____ ▶ a value that results in zero in the denominator of a rational function

_____ ▶ the quantities being multiplied in an expression involving multiplication

_____ ▶ the result of a multiplication expression

New Vocabulary Fill in each blank with the correct term or phrase.

least common ▶ To add or subtract fractions with unlike _____,
denominator (LCD) you need to rename _____ using the least common multiple of the denominators, called the least common denominator (LCD).

least common ▶ The least common multiple (LCM) is the least _____
multiple (LCM) that is a _____ of two or more numbers or polynomials.

Lesson 11-6 (continued)

Main Idea	Details
Add and Subtract Rational Expressions with Like Denominators	Find each sum. 1. $\dfrac{3k}{k-2} + \dfrac{4}{k-2}$ 2. $\dfrac{4n}{2n+5} + \dfrac{3n}{2n+5}$ _____ _____
Add and Subtract Rational Expressions with Unlike Denominators	Complete the table below to illustrate the steps involved in subtracting rational expressions with unlike denominators.

Add or Subtract Rational Expressions with Unlike Denominators	
Step 1	
Step 2	
Step 3	
Step 4	

Helping You Remember How can you use what you know about addition and subtraction of rational numbers that have like denominators to remember how to add and subtract rational expressions that have like denominators?

_____

_____

_____

11-7 Mixed Expressions and Complex Fractions

What You'll Learn

Scan the text in this lesson. Write two facts you learned about mixed expressions and complex fractions as you scanned the text.

1. _____
_____
_____
_____

2. _____
_____
_____
_____

Active Vocabulary

New Vocabulary Label the expressions with the correct terms. Write *mixed expression* or *complex fraction* in each blank.

complex fraction

mixed expression

$5 - \dfrac{2}{x-1}$ _____

$2 + \dfrac{3}{x} - \dfrac{2}{x^2}$ _____

$\dfrac{x-1}{2 - \dfrac{5}{x+2}}$ _____

$\dfrac{\dfrac{2x^2y}{5xy^3}}{\dfrac{3x^5y^3}{4xy}}$ _____

Lesson 11-7 (continued)

Main Idea	Details
Simplify Mixed Expressions	Write $4 + \dfrac{5}{x+3}$ as a rational expression. Show your work.
Simplify Complex Fractions	Follow the steps below to simplify $\dfrac{\frac{c^2 d^3}{b^4}}{\frac{c^4 d^2}{b^3}}$.

Step 1: Write the complex fraction as a division expression. _____

Step 2: Multiply by the reciprocal. _____

Step 3: Divide out common factors and simplify. _____

Helping You Remember

Describe an easy way to recognize a mixed expression.

_____

_____

_____

11-8 Rational Equations

What You'll Learn Skim the Examples in the lesson. Predict two things you think you will learn about rational equations.

1. _____

    _____

    _____

2. _____

    _____

    _____

Active Vocabulary **New Vocabulary** Fill in each blank with the correct term or phrase.

extraneous solution ▶ When a solution of a rational equation results in _____ in the _____, that solution must be excluded. Such solutions are called extraneous solutions.

rational equation ▶ A rational equation contains one or more rational _____.

rate problem ▶ Rational equations can be used to solve rate problems, such as problems involving _____.

work problem ▶ You can use _____ to solve work problems, or problems involving work rates.

NAME _____ DATE _____ PERIOD _____

Lesson 11-8 (continued)

Main Idea	Details
Solve Rational Equations	Solve the rational equation $\dfrac{4}{x-3} = \dfrac{6}{x-2}$. Check the solution.
Use Rational Equations to Solve Problems	Raymond can mow a lawn in 45 minutes using a push mower. Alex can mow the same lawn in 20 minutes using a riding mower. How long would it take them to mow the lawn working together? Raymond: Alex: **Step 1:** Find the portion of the job each person does in 1 minute. **Step 2:** Write a rational equation. **Step 3:** Solve for t. Round to the nearest tenth of a minute.

Helping You Remember A good way to remember an approach to a mathematical concept is to associate a word with it. Think of a word that can help you remember that multiplying by the LCD is one method you can use to solve a rational equation.

_____

_____

NAME _____ DATE _____ PERIOD _____

 Chapter 11 Rational Functions and Equations

Tie It Together

Fill in the graphic organizer. Add details if space permits.

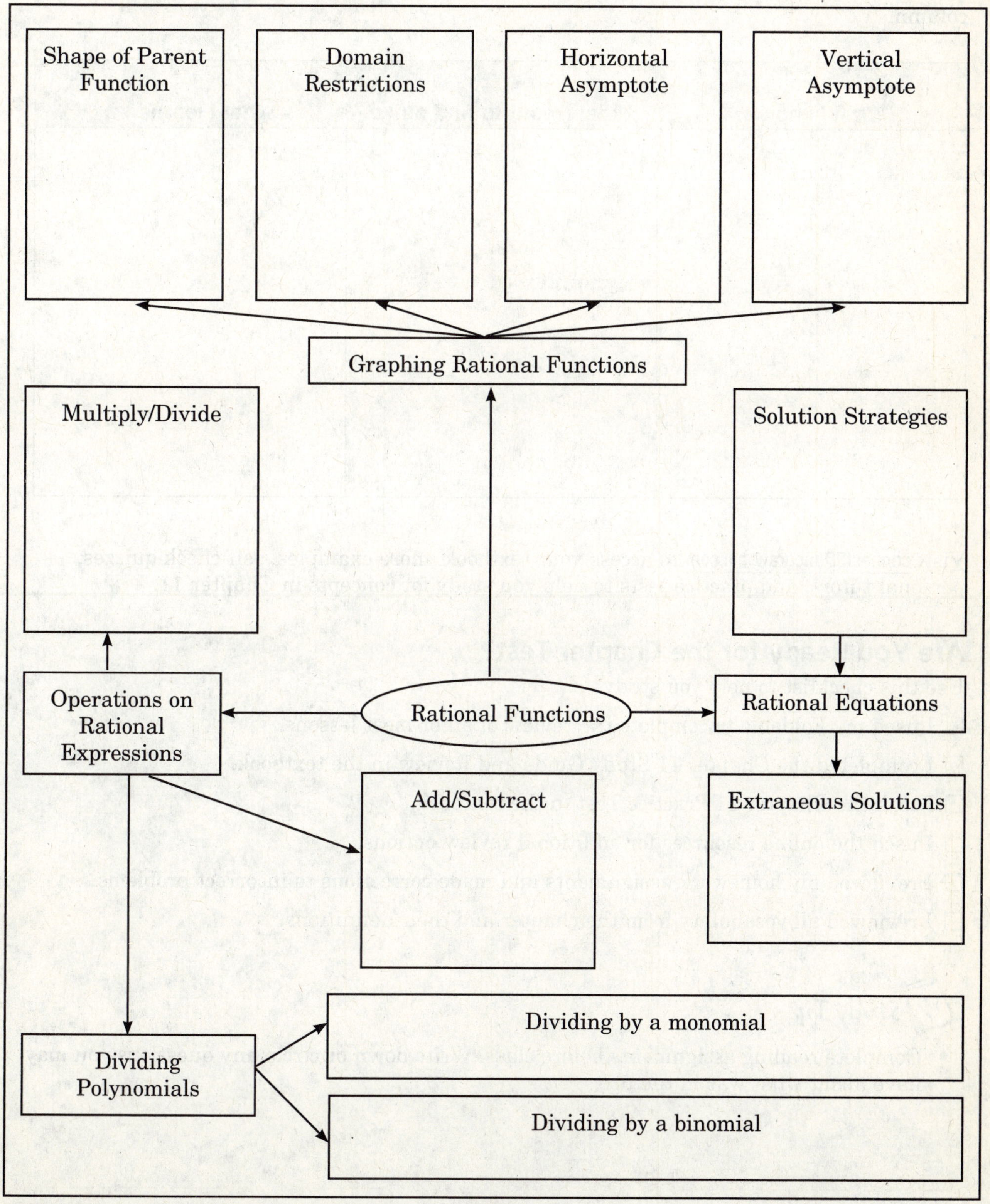

- Shape of Parent Function
- Domain Restrictions
- Horizontal Asymptote
- Vertical Asymptote

→ Graphing Rational Functions

- Multiply/Divide
- Solution Strategies

Operations on Rational Expressions ← Rational Functions → Rational Equations → Extraneous Solutions

- Add/Subtract

Dividing Polynomials
- Dividing by a monomial
- Dividing by a binomial

NAME _____ DATE _____ PERIOD _____

Rational Functions and Equations

Before the Test

Review the ideas you listed in the table at the beginning of the chapter. Cross out any incorrect information in the first column. Then complete the table by filling in the third column.

K What I know...	W What I want to find out...	L What I learned...

Visit connectED.mcgraw-hill.com to access your textbook, more examples, self-check quizzes, personal tutors, and practice tests to help you study for concepts in Chapter 11.

Are You Ready for the Chapter Test?

Use this checklist to help you study.

☐ I used my Foldable to complete the review of all or most lessons.

☐ I completed the Chapter 11 Study Guide and Review in the textbook.

☐ I took the Chapter 11 Practice Test in the textbook.

☐ I used the online resources for additional review options.

☐ I reviewed my homework assignments and made corrections to incorrect problems.

☐ I reviewed all vocabulary from the chapter and their definitions.

- Complete reading assignments before class. Write down or circle any questions you may have about what was in the text.

NAME _____ DATE _____ PERIOD _____

Statistics and Probability

Before You Read

Before you read the chapter, respond to these statements.
1. Write an **A** if you agree with the statement.
2. Write a **D** if you disagree with the statement.

Before You Read	Statistics and Probability
	• Unbiased surveys are random.
	• The sum of the probabilities for all values of x is 1.
	• A graph that shows a cluster of data about the mean is an average distribution.
	• Theoretical probability is the ratio of the frequency of an outcome to the total number of events or trials.
	• In a permutation, order matters; but in a combination, order does not matter.

FOLDABLES Study Organizer Construct the Foldable as directed at the beginning of this chapter.

Note Taking Tips

- When you take notes, it is often a good idea to use symbols to emphasize important concepts.
- When taking notes, make annotations.
  Annotations are usually notes taken in the margins of books you own to organize the text for review or study.

Chapter 12   205   Glencoe Algebra 1

NAME _____ DATE _____ PERIOD _____

CHAPTER 12 Statistics and Probability

Key Points

Scan the pages in the chapter and write at least one specific fact concerning each lesson. For example, in the lesson on statistics and parameters, one fact might be that a parameter is a measure that describes a characteristic of a population. After completing the chapter, you can use this table to review for your chapter test.

Lesson	Fact
12-1 Samples and Studies	
12-2 Statistics and Parameters	
12-3 Distributions of Data	
12-4 Comparing Sets of Data	
12-5 Simulation	
12-6 Permutations and Combinations	
12-7 Probability of Compound Events	
12-8 Probability Distributions	

NAME _____ DATE _____ PERIOD _____

12-1 Samples and Studies

What You'll Learn Scan the lesson. List two headings you would use to make an outline of this lesson.

1. _____

2. _____

Active Vocabulary **New Vocabulary** Write the correct term next to each definition.

_____ ▶ a sample that favors one group over another

_____ ▶ a method of data collection in which a process is implemented and responses are studied

_____ ▶ a method of data collection in which a sample is observed for certain patterns or behaviors

_____ ▶ the entire group about which conclusions are to be drawn

_____ ▶ a portion of a larger group

_____ ▶ a sample that is equally likely to be chosen as any other sample from the population

_____ ▶ a method of data collection in which responses are gathered from a sample of the population

Vocabulary Link *Bias* is a word that is used in everyday English. Find the definition of *bias* using a dictionary. Write how the definition of *bias* can help you remember the mathematical definition of *biased sample*.

_____

_____

_____

_____

NAME _____ DATE _____ PERIOD _____

Lesson 12-1 *(continued)*

Main Idea	Details
Sampling	A supermarket manager is interested in finding out whether or not shoppers would like an exotic foods section. He distributes 500 questionnaires to people shopping in the store. **a.** Identify the sample, and determine the population from which it was selected. _____ _____ **b.** Classify the type of sample as *simple, systematic, self-selected, convenience,* or *stratified*. _____
Studies	Complete the definition of each study type listed in the following table.

Study Types	
Type	**Definition**
survey	
observational study	
experiment	

Helping You Remember To remember what a stratified random sample is, look up the word *stratified* in a dictionary. What everyday meaning do you find that seems closest to the mathematical meaning presented in this lesson?

_____

_____

Chapter 12   **208**   Glencoe Algebra 1

NAME _____ DATE _____ PERIOD _____

12-2 Statistics and Parameters

What You'll Learn Scan the text under the *Now* heading. List two things you will learn about in the lesson.

1. _____

2. _____

_____

Active Vocabulary **Review Vocabulary** Write the correct term next to each definition. *(Lesson 0-12)*

_____ ▶ describe how widely data values vary

_____ ▶ measures of what is average

_____ ▶ the sum of the values in a data set divided by the total number of values in the set

New Vocabulary Write the correct term next to each definition.

_____ ▶ the average of the absolute values of the differences between the mean and each value in the data set

_____ ▶ the range, quartiles, and interquartile range

_____ ▶ a measure that describes a characteristic of the population

_____ ▶ a measure that describes a characteristic of the sample

_____ ▶ using the statistics of a sample to draw conclusions about the entire population

_____ ▶ the square of the standard deviation

Chapter 12   209   Glencoe Algebra 1

NAME _____ DATE _____ PERIOD _____

Lesson 12-2 (continued)

Main Idea	Details
Statistics and Parameters	A random sample of 500 pet owners in the United States is surveyed about the number of times they visit the veterinarian each year. The mean number of visits is calculated. Identify the sample and the population. Then describe the sample statistic and the population parameter. Sample: _____ Population: _____ Statistic: _____ Parameter: _____
Statistical Analysis	Follow the steps below to find the standard deviation of the data set {5, 7, 8, 10, 5}. Step 1: Find the mean of the data set. _____ Step 2: Find the variance of the data. _____ Step 3: Take the square root of the variance. _____

Helping You Remember

A good way to remember a mathematical concept is to explain it to somebody else. Suppose a classmate is having difficulty distinguishing between variance and mean absolute deviation. Explain the difference to him and give an example of each.

_____

_____

_____

12-3 Distributions of Data

What You'll Learn Skim the Examples in this lesson. Predict two things you think you will learn about distributions.

1. _____

_____

2. _____

_____

Active Vocabulary **New Vocabulary** Match the term with its graph by drawing a line to connect the two.

negatively skewed distribution

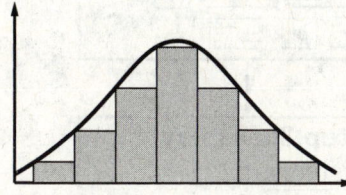

symmetric distribution

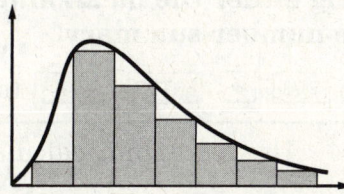

positively skewed distribution

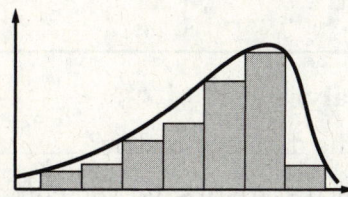

NAME _____ DATE _____ PERIOD _____

Lesson 12-3 (continued)

Main Idea	Details
Describing Distributions	Use a graphing calculator to construct a histogram for the data, and use it to describe the shape of the distribution. 6, 3, 5, 7, 5, 6, 5, 1, 6, 4, 7, 7, 4, 7

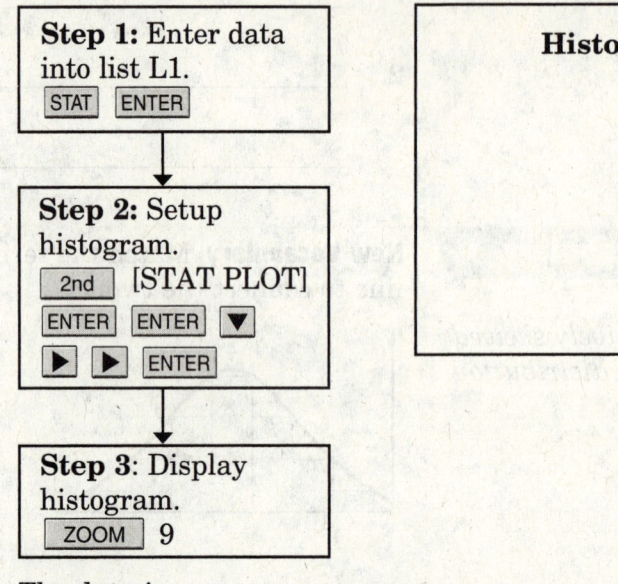

The data is _____.

Analyzing Distributions

Describe the center and spread of the data above using either the mean and standard deviation or the five-number summary.

Use [STAT] [▶] [ENTER] [ENTER] to display the **1-Var Stats**.

Use a graphing calculator to describe the data.

Analyze

Is the data symmetric? _____

Which statistics did you choose? _____

Why? _____

Describe the data: _____

_____

Chapter 12 212 Glencoe Algebra 1

NAME _____ DATE _____ PERIOD _____

12-4 Comparing Sets of Data

What You'll Learn

Scan the lesson. Write two facts you learned as you scanned the text.

1. _____

_____

2. _____

_____

_____

Active Vocabulary

Review Vocabulary Fill in each blank with the correct term or phrase *(Lesson 0–12)*.

▶ *measure of central tendency*

Data can be summarized according to three measures of central tendency: the _____, the _____, and the _____.

Vocabulary Link Explain the meaning of linear transformation in your own words. Describe how the mean, median, mode, range, and standard deviation are effected by a transformation.

_____

_____

_____

_____

_____

_____

Lesson 12-4 (continued)

Main Idea	Details
Transformations of Data	Find the mean, median, mode, range, and standard deviation of the data set obtained after adding 6 to each value. {8, 12, 9, 17, 13, 11, 12, 10, 7} New data values: _____ Mean _____ Mode _____ Standard Deviation _____ Median _____ Range _____
Comparing Distributions	Consider the points scored by the Cats football team and each of their opponents. **Cats:** 14, 20, 24, 12, 33, 18, 9, 38, 27 **Opponents:** 18, 35, 15, 24, 3, 10, 16, 21, 28 a. Describe the shape of each distribution. Cats _____ Opponents _____ b. Compare the data sets using either the means and standard deviations or the five-number summaries. Justify your choice. The Cats' scores are _____, so use the _____ to compare the data sets. Description: _____

Chapter 12   214   Glencoe Algebra 1

12-5 Simulations

What You'll Learn

Scan the text in the lesson. Write two facts you learned about probability and simulations as you scanned the text.

1. _____
_____
_____
_____

2. _____
_____
_____

Active Vocabulary

New Vocabulary Write the definition next to each term.

experimental probability ▶ _____
_____

relative frequency ▶ _____
_____

simulation ▶ _____
_____

theoretical probability ▶ _____
_____

Lesson 12-5 (continued)

Main Idea	Details
Experimental Probability	Claire correctly answered 17 questions out of 20 on a multiple choice test. What is the experimental probability that she answers a question correctly? Express your answer as a percent.
Simulations	A field goal kicker on a football team typically makes 2 out of 3 field goal attempts. Describe how to simulate a field goal attempt. Perform the simulation and predict the number of field goals the kicker will make in his next 20 attempts.

Roll	Number

_____

_____

_____

_____

Helping You Remember Mathematical concepts are easier to remember if you can explain them in your own words. How would you describe the difference between theoretical probability and experimental probability?

_____

_____

NAME _____ DATE _____ PERIOD _____

12-6 Permutations and Combinations

What You'll Learn Scan the lesson. Predict two things that you expect to learn based on the headings and the Key Concept box.

1. _____
   _____

2. _____
   _____

Active Vocabulary

Review Vocabulary Write the definition next to the term. *(Lesson 0–11)*

sample space ▶ _____

New Vocabulary Write the definition next to each term.

combination ▶ _____
_____

factorial ▶ _____
_____

permutation ▶ _____
_____

Vocabulary Link *Permutation* is a word that is used in everyday English. Find the definition of *permutation* using a dictionary. Write how the definition of *permutation* can help you remember the mathematical definition of a *permutation*.

_____
_____
_____

Chapter 12   217   Glencoe Algebra 1

Lesson 12-6 (continued)

Main Idea	Details
Permutations	Tina has 4 chores to do today. Complete the following table to determine the number of different ways Tina can do her chores if she does one at a time.

Number of Permutations	Choices for 1st chore	Choices for 2nd chore	Choices for 3rd chore	Choices for 4th chore
P				
Fundamental Counting Principle				
$P =$				

There are _____ different ways Tina can do her chores if she does one at a time.

Combinations

There are 10 players on a basketball team. How many different 5-player starting lineups are possible? Show your work.

_____

Helping You Remember

To help you remember how the terms *permutation* and *combination* are different, think of everyday words that start with the letters P and C that illustrate the meaning of each term. Explain how the words illustrate the two terms.

_____

_____

_____

_____

12-7 Probability of Compound Events

What You'll Learn

Skim the lesson. Write two things you already know about probability of compound events.

1. _____

_____

_____

2. _____

_____

_____

Active Vocabulary

New Vocabulary Match the term with its definition by drawing a line to connect the two.

compound event — events in which the outcome of one event does not affect the outcome of the other event

conditional probability — an event that is made up of two or more simple events

dependent events — events that cannot occur at the same time

independent events — events in which the outcome of one event affects the outcome of the other event

mutually exclusive events — the probability that an event will occur, given that another event has already occurred

Vocabulary Link Think of the meaning of the word *dependent*. Explain how this makes sense in the context of dependent events.

_____

_____

_____

_____

Lesson 12-7 (continued)

Main Idea	Details
Independent and Dependent Events	Model the probability of two independent events by sketching a Venn diagram in the box below. 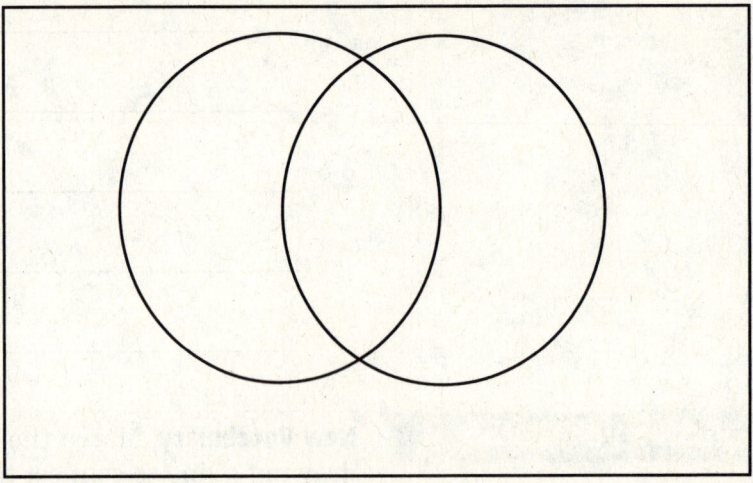
Mutually Exclusive Events	A number cube labeled 1 through 6 is rolled. Find each probability. 1. $P(1 \text{ or } 4)$ 2. $P(\text{even number})$ _____ _____

Helping You Remember

Look up the following terms in a dictionary. Write the definitions that best relate to the way these terms are used in probability.

Independent _____

Dependent _____

Exclusive _____

Inclusive _____

Chapter 12   Glencoe Algebra 1

12-8 Probability Distributions

What You'll Learn

Skim the Examples in this lesson. Predict two things you think you will learn about probability distributions.

1. _____

2. _____

Active Vocabulary

New Vocabulary Fill in each blank with the correct term or phrase.

discrete random variable ▶ A discrete random variable is a random variable with a _____ number of possibilities.

expected value ▶ Expected value is the sum of all possible values for a random variable, each value multiplied by its _____.

probability distribution ▶ A probability distribution is the probability of every possible value of the _____.

probability graph ▶ A probability graph is a graph that displays a probability _____.

random variable ▶ A random variable is a variable with a value that is the _____ of a random event.

Lesson 12-8 (continued)

Main Idea	Details
Random Variables and Probability	The table shows the grade distribution on a final exam. Find the probability that a randomly chosen student earned a B.

Grade	Number of students
A	6
B	12
C	9
D	1

Probability Distributions

Complete the following table to show the probability distribution for the number of heads when three coins are tossed. Then find the expected value.

Probability Distribution	
Number of Heads	Probability
0	
1	
2	
3	

Expected Value

$E(X) = 0 \cdot \underline{} + 1 \cdot \underline{} +$
$2 \cdot \underline{} + 3 \cdot \underline{}$

$E(X) = \underline{}$

Helping You Remember

Sometimes remembering a simple example is a good way for you to remember a more complicated mathematical concept. Show how you can use the outcomes of tossing a coin to describe how the probabilities of the possible outcomes add up to 1.